VOLUME 1

RANDOM COUNTS IN SCIENTIFIC WORK

THE PENN STATE STATISTICS SERIES

An International Series in Statistics and Applications

General Editor: G. P. Patil, Professor of Statistics
The Pennsylvania State University

This new series brings volumes of importance to statisticians and scientists in related disciplines.
The series is organized in three general categories:

I. Monographs on Statistical Distributions and Applications

II. Monographs on Related Topics in Statistics

III. Edited multi-author volumes on subjects of contemporary interest in statistics

RANDOM COUNTS IN SCIENTIFIC WORK
Expanded from the Proceedings of the
Biometric Society Symposium
Dallas, Texas, December, 1968
(American Association for the
Advancement of Science Conference)

VOLUME 1
RANDOM COUNTS IN MODELS AND STRUCTURES

G. P. PATIL, Editor
Department of Statistics
The Pennsylvania State University

The Pennsylvania State University Press
University Park and London

Library of Congress Catalogue Card Number 73-114351

Standard Book Number 271-00114-3

Printed in the United States of America

The Pennsylvania State University Press
University Park, Pennsylvania 16802

The Pennsylvania State University Press, Ltd.
London W. 1

CONTENTS

PREFACE

As Professor C. R. Rao has put it: "Endowed with a wide and rapidly expanding literature, both in diverse fields of applied science and statistics, the subject of random counts in scientific work or discrete distributions and their applications offers a fascinating area of study and research". It is of course conceivable that older and more familiar theories of statistical inference may be specifically needed or used to analyze exclusively the data on random counts. However, it is evident that new and novel mathematical and statistical approaches and procedures pertaining to special types of counted data related to individual scientific problems are long overdue.

These new procedures may be evolved in two ways. One method is to investigate in detail individual discrete distributions for their mathematical and statistical properties and for their natural applications. The second approach is to concentrate on a scientific problem that generates or requires counted data and then create and coordinate appropriate discrete models and methods so as to arrive at a meaningful solution of the scientific problem.

Random counts thus has much to offer for the advancement of the basic and applied sciences, as well as for the advancement of statistics itself.

From this premise and the multidisciplinary background of the annual conferences of the American Association for the Advancement of Science, I had felt for some time that the Biometric Society and the Statistics Section of the AAAS would provide the proper atmosphere for a much needed program on the subject of random counts in scientific work.

During the time that I served on the program committee of the Biometric Society and while I was the program chairman for 1968 meetings, I had several instructive and encouraging discussions with the immediate past program chairmen D. S. Robson and E. C. Pielou, with H. O. Hartley, Biometric Society ENAR President, and with C. I. Bliss, the Vice-President of the Statistics Section of the AAAS. A symposium on the subject of random counts was planned and organized at the Dallas Conference of the Biometric Society in December of 1968; twenty participants were on the program.

The symposium was a great success in that there was a very lively exchange and interaction between scientists from different fields and statisticians interested in the analysis of random counts. Although there seemed to be divergence of levels and diversity in the presentations, this turned out to be more of an asset than a liability.

Because of the quality and value of the presented papers, as well as widespread interest in the topic, the participants and audience felt that the proceedings should be published. It also seemed appropriate to invite additional papers from specialists who were not present in Dallas. Twenty-two papers were invited and organized in this category.

For purposes of convenience, the final total of 42 papers has been divided into three discrete subject-matter groups, published in a like number of volumes under the general title Random Counts in Scientific Work, treating random counts in (1) models and structures, (2) biomedical and social sciences, and (3) physical sciences, geoscience, and business. The present volume, Random Counts in Models and Structures, represents the first of the three volumes to be published. It is expected that more volumes will be added to this series as progress and new results in the field of random counts in scientific work come about. Any comments and suggestions from readers would be most welcome.

University Park, Pennsylvania
January 5, 1970

G. P. Patil

VOLUME 1
RANDOM COUNTS IN SCIENTIFIC WORK

CHANCE MECHANISMS GENERATING THE NEGATIVE BINOMIAL DISTRIBUTIONS

M. T. BOSWELL
G. P. PATIL
Statistics Department
The Pennsylvania State University
University Park, Pennsylvania

SUMMARY

The negative binomial distribution has been studied extensively in the past, and it has been found that data arising in various situations in many fields fits this distribution. Naturally then, one wonders about the underlying chance mechanisms, and it turns out that there are several such mechanisms which have no apparent relationship. The emphasis of this paper is on bringing together in a single notation stochastic models generating the negative binomial distribution (NBD), and the basic character of the paper is that of a systematic review and exposition of the available results on the subject. A few new results and comments given will, hopefully, clarify the structure of this important distribution.

With the exception of the important o-truncated negative binomial distribution, distributions formed by randomly stopped sums, mixtures, and other modifications of the negative binomial distribution are not discussed.

1. BACKGROUND MATERIAL

The concept of the expectation of a random variable, r.v., is basic for this paper. An r.v. X can be thought of as a numerical representation of the outcome of a random experiment; and its distribution, $f_X(x)$, is $p[X=x]$ if X is discrete or its density function if X is continuous.

Definition 1.1. The probability-generating function (pgf) of a non-negative, integer-valued r.v. X is $G_X(s) \equiv E(s^X) = \sum_x s^x p\ X=x$.

Theorem 1.1. The pgf of X, a non-negative, integer-valued r.v., exists for $|s| \leq 1$ and uniquely determines the distribution of X.
Proof. See [10], p. 125.

In Table 1 we list distributions and pgf's we will need in the main body of this paper.

Theorem 1.2. Let X and Y be r.v. Then $E[E(X|Y)] = E(X)$.
Proof. See [9], p. 223.

Theorem 1.3. (Random Sum). Let $\{X_n\}$ be a sequence of independent identically distributed r.v.'s, and let N be a non-negative, integer-valued r.v. independent of the X's. Then the pgf of the random sum $Y = X_1 + \ldots + X_N$ is given by

$$G_Y(s) = G_N[G_{X_1}(s)].$$

Proof. Applying Theorem 1.2 with conditioning on N and using independence gives the result.

Theorem 1.4 (Mixtures). Suppose the conditional distribution (and therefore the pgf) of an r.v. X, given an r.v. Λ and the distribution of Λ, are known. Then the pgf of X is given by $G_X(s) = E[G_X(s)|\Lambda)]$.
Proof. The conclusion follows from Theorem 1.2 by conditioning on Λ. In this case, X is said to be a mixture of the distribution $f_X(x|\Lambda)$ with the distribution $f_\Lambda(\lambda)$, and Λ is thought of as a parameter of the distribution of X, i.e., of $f_X(x|\Lambda)$.

Theorem 1.5. Let $N(R)$, the number of "things" of a certain type located in a region R, have a Poisson distribution with parameter $\lambda A(R)$; $A(R)$ is the length of R if R is an interval or the area of R if R is a two-dimensional region. Then the location of the "things" in R, given their number $N(R)$, has the same distribution as a random sample of size $N(R)$ from the uniform distribution on R.

Proof. See [14], p. 183.

Definition 1.2. A birth and death process is a collection X(t) of non-negative, integer-valued r.v.'s (we think of X(t) as the population size at time t) satisfying as $h \to 0$ from the right:

1. $p[X(t+h)=x+1|X(t)=x] = \lambda_x(t)h+o(h)$
2. $p[X(t+h)=x-1|X(t)=x] = \mu_x(t)h+o(h)$
3. $p[X(t+h)=x|X(t)=x] = 1-[\lambda_x(t)+\mu_x(t)]h+o(h)$

The general birth and death process defined above implies the following system of difference differential equations.

$$\frac{dp_x(t)}{dt} \equiv \frac{dp[X(t)=x]}{dt} = \lambda_{x-1}(t)p_{x-1}(t)-[\lambda_x(t)+\mu_x(t)] + \mu_{x+1}(t)p_{x+1}(t),\ x = 0,1,\ldots, \text{ with } p_{-1}(t) \equiv 0. \tag{1.1}$$

This system is extremely hard to solve in its generality, and only special cases (or forms of $\lambda_x(t)$ and $\mu_x(t)$) have been treated. The special cases where $\lambda_x(t)$ and $\mu_x(t)$ do not depend on t or on x have been extensively studied in the literature. Observe that $\lambda_x(t)$ and $\mu_x(t)$ are the (infinitesimal) birth and death rates, respectively, and that if $\mu_x(t) \equiv 0$, then the process is a pure birth process.

If (instead of the birth and death process defined above) we consider the population size (or wealth of a gambler) only at discrete time points like every hour or day (or after every bet in a gambling situation) and if we allow the population size to change by an arbitrary amount, then we have a Markov chain.

Definition 1.3. A sequence $\{X_n\}$ of non-negative, integer-valued r.v.'s is called a stationary Markov chain if the transition probabilities $p_{ij} = p[X_{n+1}=j|X_n=i]$ do not depend on n. The transition probabilities are usually displayed in matrix form, $\underset{\sim}{A} = (p_{ij})$. We say the system is in state j at step n if $X_n = j$.

Markov chains have been studied extensively. We list below the definitions and results needed in this paper. These results can be found in many books (cf., for example, [9]).

Definition 1.4. A Markov chain is irreducible if and only if it is possible to get from every state to every other state with positive probability in a finite number of steps.

Definition 1.5. The j-th state has period $m>1$ if and only if the probability of going from state j to itself in n steps is zero (unless n is a multiple of m). Furthermore, the j-th state is <u>aperiodic</u> if and only if no such n exists.

<u>Definition 1.6.</u> A state is <u>persistent</u> if any only if the probability of ever returning is one.

<u>Definition 1.7.</u> An aperiodic persistent state is <u>ergodic</u> if and only if it has finite expected return time.

<u>Theorem 1.6.</u> If a Markov chain is irreducible, then all states are of the same type (aperiodic, persistent, etc.).

<u>Theorem 1.7.</u> For a Markov chain with one step transition probabilities A, the n-step transition probabilities are given by $\underset{\sim}{A}^n$, where the probabilities of going from one state to another in n steps are called the n-step transition probabilities.

<u>Theorem 1.8.</u> In an irreducible ergodic chain the n-step transition probabilities converge to a limit which is independent of the initial state. That is,

$$\lim_{n\to\infty} A^n = B \equiv (b_{ij}), \tag{1.2}$$

where the rows of B are identical. Furthermore, $b_j \equiv b_{ij}$ for all i, b_j is positive, and $\Sigma b_j = 1$.

<u>Proof.</u> See [9], p. 393.

<u>Theorem 1.9 (Continuity Theorem).</u> Suppose that for every fixed n the sequence $a_{0,n}, a_{1,n}, \ldots$ is a probability distribution. In order that a limit, $a_i = \lim a_{i,n}$ as $n\to\infty$, exist for every $i\geq 0$, it is necessary and sufficient that the limit $A(s) = \lim \sum_{i=0}^{\infty} a_{i,n} s^i$ as $n\to\infty$, exist for each s in the open interval $0<s<1$, in which case $A(s) = \sum_{i=0}^{\infty} \left(a_i s^i\right)$.

<u>Proof.</u> See [9], p. 280.

2. NBD <u>AS A WAITING TIME DISTRIBUTION</u>

Consider a sequence of Bernoulli trials with probability p of a success. Let X be the number of failures before the first success and Y be the number of failures before the k-th success. Then X has the geometric distribution $f_X(x) = q^x p$, $x = 0,1,\ldots$, with <u>pgf</u> $G_X(s) = p/(1-qs)$. Since Y is the sum of k independent geometric r.v.'s, then $G_Y(s) = [p/(1-qs)]^k$. Therefore Y has a negative binomial distribution with parameters k and p.

If we let Z be the number of trials to get k successes, then $Z = Y+r$, and $G_Z(s) \equiv E(s^{Y+k}) = s^k G_Y(s)$. Therefore $G_Z(s) = [sp/(1-qs)]^k$, which is the pgf of the translated negative binomial distribution or Pascal distribution. Of course Z is called the waiting time to the k-th success. Some references are [9], p. 164; [20], p. 179.

3. NBD AS A POISSON (RANDOM) SUM OF LOGARITHMIC r.v.'s

Let $\{X_n\}$ be a sequence of independent, identically distributed logarithmic r.v's with parameter θ, let N be independently distributed Poisson with parameter λ, and let $Z = X_1+\ldots+X_N$. Then by Theorem 1.3,

$$G_Z(s) = G_N\left[G_{X_1}(s)\right] = \exp\left[\lambda\frac{\ln(1-\theta s)}{\ln(1-\theta)}-1\right]$$

$$= \exp\left[\ln\left(\frac{1-\theta s}{1-\theta}\right)\lambda/\ln(1-\theta)\right]$$

$$= [(1-\theta)/(1-\theta s)]^{-\lambda/\ln(1-\theta)}.$$

Therefore by Theorem 1.1, Z has a negative binomial distribution with parameters $k = -\lambda/\ln(1-\theta) > 0$ and $p = 1-\theta$. Some references are [13], [16], and [22].

4. NBD AS A POISSON MIXTURE WITH THE GAMMA DISTRIBUTION

Let X (given Λ) have a Poisson distribution with parameter Λ where Λ is itself a r.v. having the gamma distribution with parameters θ and ξ. Then by Theorem 1.4,

$$G_X(s) = E\left[G_X(s|\Lambda)\right] = E\{\exp[\Lambda(s-1)]\}.$$

$$= \int_0^\infty \frac{\theta^\xi x^{\xi-1} e^{x(s-1-\theta)}}{\Gamma(\xi)}dx$$

$$= \left(\frac{\theta}{\theta+1-s}\right)^\xi \int_0^\infty \frac{(\theta+1-s)^\xi x^{\xi-1} e^{-(\theta+1-s)x}}{\xi}dx$$

$$= [\theta/(\theta+1-s)]^\xi$$

$$= \left[\frac{\theta}{1+\theta}/(1-\frac{1}{1+\theta}s)\right]^\xi.$$

Therefore X has a negative binomial distribution with parameters $k = \xi$ and $p = \theta/(1+\theta)$. Some references are [3], p. 8; [12],

p. 39; and [21], p. 18.

5. NBD AS A LIMIT OF THE POLYA DISTRIBUTION

Polya's urn scheme is well known (cf. [10], p. 136). One samples from an urn containing N balls with a fraction p white balls and a fraction 1-p black balls. After each trial the ball drawn is replaced along with $c \equiv \beta N$ others. Let X be the number of white balls drawn in n trials. The distribution of X is given by

$$f_X(x) = \binom{n}{x}\frac{p(p+\beta)\ldots[p+(x-1)\beta]q(q+\beta)\ldots[q+(n-x-1)\beta]}{1(1+\beta)\ldots[1+(n-1)\beta]}, \tag{5.1}$$

for x = 0,1,...,n. This distribution is called the Polya distribution.

Suppose a ball is drawn regularly so that n balls are drawn in time t (e.g, one draws a ball every t/n min). Taking the limit as $n\to\infty$, if we hold the parameters constant, would yield an infinite number of white balls drawn.

Theorem 5.1. Taking the limit of the Polya distribution, (5.1), as $p \to 0$, $\beta \to 0$, $n \to \infty$, so that $np \to t$ and $\beta n \to bt$ for some constant b, gives the negative binomial distribution with parameters $k = 1/b$ and $p = 1/(1+bt)$.

NOTE: This theorem is often stated in a form equivalent to the above; however, in the proof, as far as we know, it is always assumed that p/β is an integer. However p/β converges to $1/b$; therefore $1/b$ is an integer, too. With this assumption the proof is straightforward but lengthy (cf. [4]) and is omitted here. Some other references are [9], p. 480; and [18].

6. NBD AS A LIMIT OF A BINOMIAL MIXTURE WITH THE β-DISTRIBUTION

The limiting process here is the same as that of Section 5 above. We obtain the Polya distribution as a mixture (cf. [21] p. 29, #28).

Theorem 6.1. Let X have a binomial distribution with parameters n and P, where P is a random variable having the β-distribution with parameters p/β and $(1-p)/\beta$. Then the unconditional distribution of X is the Polya distribution, (5.1).

Proof. Now $f_X(x) = \int_0^1 f_X(x|P=u)f_P(u)du$

$$= \binom{n}{x}\frac{\Gamma(1/\beta)}{\Gamma(p/\beta)\Gamma[(1-p)/\beta]}\int_o^1 u^{x+p/\beta-1}(1-u)^{n-x+(1-p)/\beta-1}dn$$

$$= \binom{n}{x}\frac{\Gamma(1/\beta)\Gamma(x+p/\beta)\Gamma[n-x+(1-p)/\beta]}{\Gamma(p/\beta)\Gamma[(1-p)/\beta]\ (n+1/\ )}$$

The conclusion follows from the fact that $\Gamma(m+1) = m\Gamma(m)$.

7. NBD AS THE POLYA PROCESS

Consider a pure birth process with birth rate $\lambda_x=(1+xb)/(1+bt)$. Let $p_x(t)$ be the probability of x births in time 0 to t. From (1.1)

$$\frac{dp_x(t)}{dt} = \lambda_{x-1}(t)p_{x-1}(t)-\lambda_x(t)p_x(t), \quad x = 0,1,\ldots,$$

where $p_{-1}(t) \equiv 0$. The boundary conditions are $p_x(0) = 0$ if $x > 0$ and $p_0(0) = 1$. This can be solved iteratively, or it is easy to see that the negative binomial distribution with parameters $k = 1/b$ and $p = 1/(1+bt)$ satisfies the differential equation and boundary conditions, which makes it the unique solution.

The negative binomial distribution with these parameters is actually a stochastic process. That is, for each fixed t there is a different r.v. This process was obtained in Section 5 as a limit of the Polya distribution, and in fact it is called the Polya process. We now obtain $\lambda_x(t) = (1+xb)/(1+bt)$ as a limit from the Polya distribution. As in Section 5, assume that n balls be drawn in time t. Let $p_{x,n}$ be the conditional probability of drawing a white ball on the (n+1)th draw, given that x white balls were drawn in the first n drawings. Clearly the "condition" of the urn before this draw is $pN+x\beta N$ white balls and $(1-p)N+(n-x)\beta N$ black balls. Therefore

$$p_{x,n} = \frac{pN+x\beta N}{N+n\beta N} = \frac{p+x\beta}{1+n\beta},$$

assuming for n sufficiently large or $\Delta \equiv (t/n)$ sufficiently small that the probability that two or more white balls being drawn in time Δ is $o(\Delta)$; then the average birth rate from t to $t+\Delta$ is

$$\frac{p_{x,n}}{\Delta} = \frac{1}{t}\cdot\frac{np+xn\beta}{1+n\beta} + \frac{o(\Delta)}{\Delta}.$$

Taking the limit as $n \to \infty$, $p \to 0$, and $\beta \to 0$ such that $np \to t$ and $n \to\ bt$ gives the birth rate $\lambda_x(t) = \lim(p_{x,n}/\Delta) = (1+xb)/(1+bt)$ at time t, which was to be shown. References: [18], p. 17; and [9], p. 480.

8. NBD AS A POPULATION GROWTH WITH IMMIGRATION

8.1 The Yule-Furry Process

Consider a population whose members can, by splitting or otherwise, give birth to new members but cannot die. Assume that the probability that any member in time t to t+h, gives birth to a new one is $\beta \cdot h + o(h)$ and that each member acts independently of every other member. Then the birth rate is independent of time and is proportional to the population size. That is, $\lambda_x(t) = x\beta$, if the population size at t = 0 is 1. From (1.1) it is easy to find that the population size at time t due to a single individual at t = 0 is given by

$$p_x(t) = e^{-\beta t}(1-e^{-\beta t})^{x-1}, \quad x=1,2,\ldots.$$

Let X(t) be the population size at time t due to a single individual at t = 0. Then

$$G_{X(t)}(s) = \sum_{x=1}^{\infty} s^x p_x(t) = se^{-\beta t}/\left[1-(1-e^{-\beta t})s\right].$$

Let Y(t) be the population size at time t due to n individuals at time t = 0. Then Y(t) is a Yule-Furry process. Since the individuals act independently, $G_{Y(t)}(s) = \left\{se^{-\beta t}/\left[1-(1-e^{-\beta t})s\right]\right\}^n$. If one is interested in the increase in size of the population instead of the population size, one considers Z(t) = Y(t)-n. Then

$$G_{Z(t)}(s) \equiv Es^{Y(t)-n} = s^{-n}G_{Y(t)}(s) = \left\{e^{-\beta t}/\left[1-(1-e^{-\beta t})s\right]\right\}^n,$$

which is the probability-generating function of the negative binomial distribution with parameters k = n and $p = e^{-\beta t}$. It is clear that the Yule-Furry process is a translated negative binomial process.

8.2 Growth with Immigration

Suppose individuals reproduce according to some process. In addition suppose individuals immigrate according to a Poisson process and thereafter reproduce by the same process. Assume all individuals reproduce independently. Let W(t) be the population size at time t due to all immigrants from time 0 to t, let X(t) be the size at t due to one individual present at t = 0, and let Y(t) be the population size at time t due to all sources.

Let N(t) be the number of immigrants from time 0 to t, and let $T_1, T_2, \ldots$ be the times of immigrations.
Then

$$G_{W(t)}(s) = E\left[s^R\right] = E\left\{E\left[s^R \middle| N(t), T_1, \ldots, T_{N(t)}\right]\right\},$$

by Theorem 1.2, where $R = \sum_{j=1}^{N(t)} X(t-T_j)$.

Then by Theorem 1.5,

$$G_{W(t)}(s) = E\left[s^U\right], \text{ where } U = \sum_{j=1}^{N(t)} X(t-V_j) \text{ and } V_j \text{ is independent}$$

uniform r.v. over [0, t]. And by Theorem 1.3,

$$G_{W(t)}(s) = G_{N(t)} E\left[s^{X\ t-V_1}\right] = G_{N(t)} \int_0^t (1/t) E\left[s^{X(t-\tau)}\right] d\tau ,$$

since V_1 is uniform,

$$= \exp\left\{(\lambda t)\left[(1/t)\int_0^t G_{X(t-\tau)}(s) - 1\right] d\tau\right\}, \text{ since } N(t)$$

is Poisson,

$$= \exp\left\{\lambda \int_0^t \left[G_{X(t-\tau)}(s) - 1\right] d\tau\right\}.$$

Assuming a Yule-Furry process as in Section 8.1, we have

$$G_{W(t)} = \exp\left\{\frac{\lambda}{\beta} \int_0^t \left[\frac{\beta s e^{-\beta(t-\tau)}}{1- \ 1-e^{-\beta(t-\tau)} \ s} - \beta\right] d\tau\right\}$$

$$= \left\{e^{-\beta t} / \left[1-(1-e^{-\beta t})s\right]\right\}^{\lambda/\beta}.$$

Therefore the population size at time t due to all immigrations has a negative binomial distribution with parameters $k = \lambda/\beta$ and $p = e^{-\beta t}$. Furthermore the <u>increase</u> in the size of the population due to all sources would have a negative binomial distribution with parameters $k = n+\lambda/\beta$ and $p = e^{-\beta t}$. This example is given in [14], p. 345.

9. NBD AS A LINEAR BIRTH PROCESS

Consider a pure birth process with $\lambda_x(t) = x\beta+\lambda$, $\beta > 0$, $\lambda > 0$. From (1.1) we have $P_x'(t) = [(x-1)\beta+\lambda]P_{x-1}(t) - [x\beta+\lambda]p_x(t)$. It is not hard to show that the negative binomial distribution with parameters $k = \lambda/\beta$ and $p = e^{-\beta t}$ satisfies this system of differential equations, as well as the boundary conditions $p_0(0) = 1$

and $p_x(0) = 0$ for $x > 0$.

Recall (Section 8) that the increase in population size due to all immigrants, assuming the Yule-Furry process for reproduction also has the negative binomial distribution with parameters $k = \lambda/\beta$ and $p = e^{-\beta t}$. Basic to the Yule-Furry process is independent reproduction of all individuals, and each individual has a birth rate β (Poisson process with parameter β). Also the immigration is according to a Poisson process with parameter λ. It is clear, therefore, that the birth rate should be of the form $\lambda_x(t) = x\beta+\lambda$ for any part of the process. The boundary condition $p_0(0) = 1$ implies there are no members in the population at $t = 0$. If instead we used $p_n(0) = 1$ and $p_m(0) = 0$ for $m \neq n$, then we should have the process studied in Section 8. Or if we were interested in the increase in the size of the population instead of the actual size of the population, then the birth rate would be

$$\lambda_m(t) = (m+n)\beta+\lambda = m\beta+(n\beta+\lambda),$$

and the boundary conditions would be $p_0(0) = 1$ and $p_m(0) = 0$ if $m > 0$. Using the results obtained above with λ replaced by $(n\beta+\lambda)$, we see the process has a negative binomial distribution with parameters $k = (n\beta+\lambda)/\beta \equiv n+\lambda/\beta$ and $p = e^{-\beta t}$. This, as was expected, is the distribution obtained in Section 8.

The theory of birth process of course can be applied to non-birth situations. In particular, t could be the area of a region, and a "birth" could be the finding of a new individual as the area of search was increased. Some references are [23], p. 136; and [24], p. 25.

10. NBD AS A LINEAR BIRTH AND DEATH PROCESS

Consider a birth and death process with $\lambda_x(t) = x\lambda+\beta$ and $\mu_x = x\mu$. Analogously to Section 9 we can interpret this as each individual having a birth rate of λ and a death rate of μ and where the population as a whole has a birth rate due to immigration of β. From (1.1) we have

$$P_x'(t) = [(x-1)\lambda+\beta]p_{x-1}(t)-(x\lambda+\beta+x\mu)p_x(t)+(x+1)\mu p_{x+1}(t).$$

Multiplying by s^x, summing over all x, and simplifying gives the partial differential equation

$$\frac{\partial G_{X(t)}(s)}{\partial t} = (s-1)\left[\beta G_{X(t)}(s)+(\lambda s-\mu)\frac{\partial G_{X(t)}(s)}{\partial s}\right].$$

It is not hard to show that

$$G_{X(t)}(s) \equiv \left[p(t)/\{1-[1-p(t)]s\}\right]^{\beta/\lambda},$$

where

$$p(t) = (\lambda-\mu)/\left[\lambda e^{(\lambda-\mu)t}-\mu\right] \quad \text{if } \mu \neq \lambda, \text{ and}$$

$$= 1/(1+\lambda t) \qquad \text{if } \mu = \lambda,$$

satisfies the above differential equations, with the boundary condition $X(0) = 0$. Therefore assuming we start with a population of size zero at time $t = 0$, then the population size at time t has a negative binomial distribution with parameters β/λ and p, given above.

The limit of $G_{X(t)}(s)$ as $t \to \infty$ obviously exists if and only if $\lambda < \mu$, in which case the limit is $G(s) = \{p/[1-(1-p)s]\}^{\beta/\lambda}$, where $p = 1-\lambda/\mu$. Therefore by Theorem 1.9, if $\lambda < \mu$ the equilibrium case is negative binomial with parameter $k = \beta/\lambda$. The results in this section can be found in [17], pp. 241-243. See also [7], p. 326.

11. NBD AS THE EQUILIBRIUM CASE OF A MARKOV CHAIN

Let an ergodic irreducible Markov chain have the matrix of transition probabilities $\underset{\sim}{A} = (p_{ij})$, (for definitions of this section refer to Definitions 1.3 to 1.7). Then by Theorem 1.8,

$$\lim_{n\to\infty} \underset{\sim}{A}^n = \underset{\sim}{B} \equiv b_{ij},$$

where the rows of $\underset{\sim}{B}$ are identical ($b_{ij} = b_j$ for all i) and where $\sum_{j=1}^{\infty} b_j = 1$. Since $\underset{\sim}{A}^{n+1} \equiv \underset{\sim}{A}^n\underset{\sim}{A}$ and $\underset{\sim}{A}^n$ $\underset{\sim}{B}\underset{\sim}{A}$, then $\underset{\sim}{B}\underset{\sim}{A} = \underset{\sim}{B}$. Or

$$\underset{\sim}{x}\underset{\sim}{A} = \underset{\sim}{x}, \text{ where } \underset{\sim}{x} = (b_0, b_1, \ldots). \tag{11.1}$$

Starting with a sequence $\{a_n\}$ of positive numbers, one can construct a matrix of transition probabilities as follows: let

$$A_n = \sum_{k=0}^{n} a_k, \text{ let } p_{ij} = a_j/A_i = 1,\ldots,j, \text{ and let}$$

$$p_{ij} = 0 \text{ if } j > i.$$

<u>Theorem 11.1.</u> The above construction gives

$$b_n = \frac{a_1 \ldots a_n}{A_0 \ldots A_{n-1}} \cdot b_0 \equiv c_n b_0, \tag{11.2}$$

with

$$b_0 = 1/\left(1+\sum_{n=1}^{\infty} c_n\right).$$

Proof. From (11.1) one finds $b_1 = a_1/a_0\, b_0$ and $b_{n+1}/a_{n+1} = \left(b_n/a_n\right)\left(b_{n-1}/A_n\right)$. The conclusion follows by induction and the fact that $\Sigma b_n = 1$.

Conversely if one knows $\underset{\sim}{b}$ one can think of (11.2) as defining $\{a_n\}$ (the sequence $\{a_n\}$ is not unique; indeed, any multiple of the sequence also works).

Theorem 11.2. Solving (11.2) for $\{a_n\}$ in terms of $\underset{\sim}{b}$ gives

$$a_n = \left(\frac{b_n}{b_{n-1}}\right)\prod_{i=1}^{n-1}\left(1+\frac{b_i}{b_{i-1}}\right),$$

where a_0 is taken, without loss of generality, to be 1.

Proof. From (11.2) one finds $b_n/b_{n-1} = a_n/A_{n-1}$, $n=1,2,\ldots.$ Therefore $a_n = A_{n-1}b_n/b_{n-1}$. Then $A_n - A_{n-1} = A_{n-1}b_n/b_{n-1}$, or $A_n = A_{n-1}(1+b_n/b_{n-1})$. This reduction formula applied iteratively gives $A_n = A_0 \prod_{i=1}^{n} (1+b_i/b_{i-1})$ where $A_0 \equiv a_0 = 1$. The conclusion follows from the fact that $a_n = A_n - A_{n-1}$.

One can interpret this process as a random walk where states $E_0,\ldots,E_{n+1}$ can be reached with probabilities in the ratio $a_0:a_1:\ldots:a_{n+1}$. Then the asympotic distribution of the particle taking the random walk is $\underset{\sim}{b}$.

We now answer the question as to what leads to an asymptotic distribution which is negative binomial. That is, assume

$$b_n = \binom{n+k-1}{n}p^k(1-p)^n, \text{ where } n = 0,1,\ldots$$

$$\equiv \binom{n+k-1}{n}q^n(1-q)^k,\ q = 1-p,\ n = 0,1,\ldots.$$

Solving inductively for $\{a_r\}$ by Theorem 11.2 gives $a_0 = 1$, $a_1 = \lambda\beta$

$$a_n = q(k+n-1)\cdot(1+kq)\ldots[(n-1)+(k+n-2)q]/n!,\ n = 2,3,\ldots.$$

Since $q(k+n-1) = [n+(k+n-1)q]-n$, then $a_n = c_n - c_{n-1}$, and

$$c_n = (1+kq)\ldots[n+(k+n-1)q]/n!$$

$$= (1+q)^n\binom{p+n}{n},\ P = (k-1)q/(1+q).$$

Then $A_n \equiv \sum_{k=0}^{n} a_k = c_n$, and finally the matrix of transition probabilities is given by

$$p_{ij} = \frac{c_j - c_{j-1}}{c_i}, \text{ if } f \le i$$
$$= 0 \text{ if } j > i.$$

We remark in closing that c_j/c can be interpreted as a probability. The results and example of this section can be found in [11].

12. NBD BASED ON RANDOMLY DISTRIBUTED PARENTS AND NORMALLY DISTRIBUTED PROGENY

Suppose parent plants are distributed at random over the entire plane with an average rate of λ per unit area. Further suppose progeny plants distributed about their parents with a circular normal distribution with variances σ^2, and suppose the number of progeny from a given parent satisfies a geometric distribution with parameter μ. Then, as we will see, the number of progeny due to parents on the plane in a "small" region R has approximately a negative binomial distribution.

Lemma 12.1. Let $f(z) \equiv f(x,y)$ be the circular normal density function with zero means and variances σ^2. Then

$$\int \frac{af(z)}{a+bf(z)}\,dz = 2\pi\sigma^2 \ln\left(1+\frac{b}{2\pi\sigma^2 a}\right).$$

Proof. The proof is a straightforward exercise of changing to polar coordinates.

Let $X(R_0)$ be the number of parent plants in the region R_0, let N be the number of progeny from a given parent, let $Y(r_0)$ be the number of progeny in a fixed region R due to a parent located at r_0, and let $Y(R_0)$ be the number of progeny in R due to all parents in R_0. In this notation our assumptions are:

$X(R_0)$ has a Poisson distribution with parameter λ;

N has a geometric distribution with parameter μ; and

the location of a progeny from a parent at $r_0 \equiv (x_0, y_0)$ has a density function $f(z-r_0)$, $f(z)$ defined in Lemma 12.1, $z \equiv (x,y)$.

Lemma 12.2. If R is small in the sense that the distance between any two points of R is small, then $Y(r_0)$ is a random sum of N Bernoulli r.v.'s with parameter

$$p(r_0) \equiv \int_R f(z-r_0)\,dz \approx f(r-r_0)A(R) \text{ for any } r \in R,$$

where A(R) is the area of R.

Proof. Clearly $Y(r_0)$ is the random sum mentioned above. Now

$$p(r_0) = \int_{R-r_0} f(z)dz, \text{ where } R-r_0 = \{z-r_0 : z\varepsilon R\}.$$

The conclusion follows from the fact that f(r) is approximately constant over R.

Theorem 12.1. The number of progeny, y, located in R and due to all parents, has approximately a negative binomial distribution if R is small as defined above.

Proof. By Theorem 1.2,

$$G_{Y(R_0)}(s) \equiv E\left[s^{Y(R_0)}\right] = E\left\{E\left[s^{Y(R_0)} \mid X(R_0)\right]\right\}$$

$$= E\left(E\left\{E\left[s^{Y(R_0)} \middle| r_1,\ldots,r_{X(R_0)}\right] X(R_0)\right\}\right),$$

where r_i is the location of the i-th parent in the region R_0. By Theorem 1.5, $r_1,\ldots,r_{X(R_0)}$, given $X(R_0)$, has the same distribution as $X(R_0)$ independent uniform r.v.'s over R_0. Therefore

$$G_{Y(R_0)}(s) = E\left(E\left\{\prod_{j=1}^{X(R_0)} E\left[s^{Y(r_j)}\right] \middle| X(R_0)\right\}\right)$$

$$= E\left\{E\left[s^{Y(U)}\right]^{X(R_0)}\right\}, \text{ where U is uniform over } R_0$$

$$= G_{X(R_0)}\left\{E\left[s^{Y(u)}\right]\right\}$$

$$= \exp\left[\lambda A(R_0)\left(\int_{R_0}\left\{E\left[s^{Y(z)}\right]/A(R_0)\right\} dz-1\right.\right.$$

$$\equiv \exp\left\{\lambda\int_{R_0}\left[G_{Y(z)}(s)-1\right]dz\right\}$$

However, by Theorem 1.2,

$$G_{Y(z)}(s) = E\left\{E\left[s^{Y(z)} \mid N\right]\right\}$$

$$= E\left\{\left[1+p(z)\cdot(s-1)\right]^N\right\},$$

since Y(z), given N, is binomial N, P(z).

$$= G_N\left[1+p(z)\cdot(s-1)\right]$$

$$= (1-\mu)/\{1-\mu[1+p(z)\cdot(s-1)]\}, \text{ since N geometric } \mu$$

$$\approx (1-\mu)/[1-\mu-\mu f(r-z)A(R)\cdot(s-1)] \text{ for any r in R,}$$

by Lemma 12.2. Therefore

$$G_{Y(R_0)}(s) = \exp\left(\lambda\int_{R_0}\{(1-\mu)/[(1-\mu)-\mu f(r-z)A(R)\cdot(s-1)]-1\}dz\right).$$

Passing to the limit as R_0 goes to the entire plane, we have

$$G_Y(s) = \exp\left(\lambda\int\{(1-\mu)/[(1-\mu)-\mu f(z)A(R)(s-1)]-1\}dz\right).$$

The conclusion follows from Lemma 12.1. The results of this section can be found in [14], p. 356.

13. NBD AS A QUEUEING PROCESS

Suppose individuals arrive at a queue according to a Poisson process with parameter λ, suppose a person arriving at the queue joins the queue with probability $\beta(x)$(where x is the queue length or does not join the queue), and suppose individuals leave the queue according to a Poisson process with parameter μ. The queue length then is a birth and death process with birth rate $\lambda_x = \lambda\beta(x)$ and death rate $\mu_x = \mu$ if $x > 0$, and $\mu_x = 0$ if $x = 0$. Then from (1.1) we have

$$\frac{dp_x(t)}{dt} = \lambda_{x-1}p_{x-1}(t)-(\lambda_x+\mu)p_x(t)+\mu p_{x+1}(t) \text{ if } x > 0$$

$$= -\lambda_0 p_0(t)+\mu p_1(t) \text{ if } x = 0.$$

Taking the sum over $x = 0,1,\ldots,n$ gives

$$\frac{d\sum_{x=0}^{n} p_x(t)}{dt} = p_{n+1}(t)-\lambda_n p_n(t).$$

Assuming equilibrium is reached and taking the limit as $t \to \infty$, we find that

$$\frac{p_{n+1}}{p_n} = \frac{\lambda_n}{\mu} \equiv \frac{\lambda}{\mu}\beta(n),$$

where $p_n = \lim p_n(t)$ as $t \to \infty$. Now observe that the ratio of $p[X=n+1]/p[X=n]$ is $(1-p)(n+k)/(n+1)$, where X has a negative binomial distribution with parameters k and p. Therefore, if $\beta(x) = (x+k)/(x+1)$, then the queue length at equilibrium has a negative binomial distribution with parameters k, $(0<k<1$ so that $\beta(x)$ is a probability) and $p = (1-\lambda/\mu)$.

14. ZERO-TRUNCATED NBD AS A GROUP-SIZE DISTRIBUTION

Suppose in a certain population of N_0 individuals, groups of individuals form and reform from time to time. Assume the way a group changes is by either adding or losing a single individual. Further assume that in a short time interval the average number of groups of a certain size has a rate of change which is a linear function of the number of individuals and the number of groups directly involved plus an error term whose expected value is zero.

Let $N_x(t)$ be the number of groups at time t of size x, and let N(t) be the number of groups. Then $N_0 = \sum_{x=1}^{\infty} xN_x(t)$, and $N(t) = \sum_{x=1}^{\infty} xN_x(t)$. Of course $N_x(t) \equiv 0$ if $x > N_0$. Let $m_x(t) = E[N_x(t)]$, $x = 1,2,\ldots$; and let $m(t) = E[N(t)]$. Observe that

$$N_0 = E\left(N_0\right) = \sum_{x=0}^{\infty} xm_x(t). \tag{14.1}$$

With this notation our assumptions become

$$\begin{aligned} \underset{\sim}{D}\left[N_x(t)\right] &\equiv \left[N_x(t+h)-N_x(t)\right]/h \\ &= [a+b(x-1)]N_{x-1}(t)+[c+d(x+1)]N_{x+1}(t) - \\ &\quad [(a+bx)+(c+dx)]N_x(t)+e(h), \end{aligned}$$

where a is the group factor, b is the individual factor for attracting another individual to a group, c is the group factor, d is the individual factor for repelling an individual from a group, and e(h) is the error term which satisfies $E[e(h)]=0$. Taking expectations gives

$$\underset{\sim}{D}[m_x(t)] = [a+b(x-1)]m_{x-1}(t)+[c+d(x+1)]m_{x+1}(t)-[a+bx+c+dx]m_x(t). \tag{14.2}$$

From (14.1) and (14.2) we have

$$\underset{\sim}{D}[m(t)-m_1(t)] = (a+b)m_1(t)-(c+2d)m_2(t)$$

and

$$\underset{\sim}{D}(N_0) \equiv 0 = \sum_{x=1}^{\infty} x\underset{\sim}{D}[m_x(t)]. \tag{14.3}$$

Assuming the system reaches equilibrium at $t = \infty$, then $m_x(t) \to m_x$ and $\underset{\sim}{D}[m_x(t)] \to 0$ as $t \to \infty$ for $x = 1,2,\ldots$. Taking the limit on (14.2) and (14.3) and solving iteratively gives $m_{x+1}/m_x = (a+bx)/[c+d(x+1)]$ for $x = 1,2,\ldots$. Without loss of generality assume $d = 1$ (or multiply the numerator and denominator by d^{-1}).

Now the probability that a group selected at random (at equilibrium) is of size x is the fraction of groups of size x, $f(s) = M_x/m$. Therefore

$$\frac{f(x+1)}{f(x)} = \frac{a+bx}{c+x+1} = b\frac{(a/b)+x}{c+1+x}.$$

As at the end of Section 13 we see that if c is zero, then the group size at equilibrium has a negative binomial distribution with parameters $k = a/b$ and $p = 1-b$.

The assumptions for this model differ from those of Cohen [6]; however, the model and its development are the same. See also [7], p. 365.

15. ZERO-TRUNCATED NBD AS A ZERO-TRUNCATED POISSON MIXTURE

A random variable possessing a Poisson distribution with parameter Λ for which the 0-th case is unobservable or lost has the zero-truncated Poisson distribution

$$f(x|\Lambda) = \Lambda^x e^{-\Lambda}/(x!)(1-e^{-\Lambda}).$$

Let X be such an r.v., and further, assume that Λ is itself a random variable with distribution given by

$$f_\Lambda(\lambda) = c(1-e^{-\lambda})e^{-\theta\lambda}\lambda^{\xi-1}, \quad 0<\lambda<\infty,$$

where c is the normalizing factor given by $1/c = \frac{\Gamma(\xi)}{\theta^\xi}\left[1+(\frac{\theta}{1+\theta})^\xi\right]$. Then the distribution of X is

$$\begin{aligned} f_X(x) &= \int_0^\infty f_X(x|\lambda)f_\Lambda(\lambda)d\lambda \\ &= (c/x!)\int_0^\infty \lambda^{x+\xi-1}e^{-(1+\theta)\lambda}d\lambda \\ &= (c/x!)\Gamma(x+\xi)/(1+\theta)^{x+\xi} \\ &= \frac{\binom{x+\xi-1}{x}\left(\frac{\theta}{1+\theta}\right)\left(\frac{1}{1+\theta}\right)^x}{1-\left(\frac{\theta}{1+\theta}\right)^\xi}, \end{aligned}$$

which one recognizes as the zero-truncated negative binomial distribution, with parameters $k = \xi$ and $p = \frac{\theta}{1+\theta}$. We observe that the mixing distribution here is not, as in the case for the NBD arising as a Poisson mixture, a gamma distribution.

Table 1. Distributions and pgf's Used in This Paper

Name of Distribution	Distribution $f_X(x)$	pgf	Conditions on Parameters
uniform (two-dimensional)	$1/A(R)$ if X is in R	-	$A(R)$ = area of $R > 0$
binomial	$\binom{n}{x} p^x q^{n-x}$, $x = 0,1,\ldots,n$	$(q+ps)^n$	$0<p<1$, $q = 1-p$, $n = 1,2,\ldots$
negative binomial	$\binom{-k}{x} p^k (-q)^x$, $x = 0,1,\ldots$	$(p/(1-qs))^k$	$0<p<1, q = 1-p$, $k > 0$
logarthmic	$\alpha_\theta{}^x/x$, $x = 1,2,\ldots$	$\alpha \ln(1-\theta s)$	$0<\theta<1$, $\alpha = \frac{1}{-\ln(1-\theta)}$
Poisson	$\lambda^x e^{-\lambda}/(x!)$, $x = 0,1,\ldots$	$\exp[\lambda(s-1)]$	$\lambda > 0$
geometric (negative binomial with k=1)	$q^x p$, $x = 0,1,\ldots$	$p/(1-qs)$	$0>p>1$, $q = 1-p$
gamma	$\theta^\xi x^{\xi-1} e^{-\theta x}/\Gamma(\xi)$, $x > 0$	-	$\theta > 0$, $\lambda > 0$
beta	$\frac{\Gamma(p+q)}{\Gamma(p)\,\Gamma(q)} x^{p-1}(1-x)^{q-1}$, $0<x<1$	-	$p > 0$, $q > 0$
circular normal	$\frac{1}{2\pi\sigma^2} e^{-(1/2\sigma^2)\left[(x-\mu_1)^2+(y-\mu_2)^2\right]}$	-	$\sigma^2 > 0$, $-\infty<\mu_i<\infty$, $i = 1,2$

REFERENCES

[1] Arley, Niels. 1948. Stochastic Processes and Cosmic Radiation. Wiley, New York.

[2] Barlett, M. S. 1960. Stochastic Population Models. Methuen's monographs on applied probability and statistics. Wiley, New York.

[3] Bharucha-Reid, A. T. 1960. Elements of the Theory of Markov Processes and Their Applications. McGraw-Hill, New York.

[4] Bosch, L. 1963. The Polya distribution. Statistica Neerlandica 17:201-13.

[5] Brass, W. 1958. The distribution of births in human populations. Population Studies 12:51-72.

[6] Cohen, J. E. 1968. Some theory of the formation of groups. Personal communication.

[7] Coleman, J. S. 1964. Introduction to Mathematical Sociology. Collier-Macmillan, London.

[8] Dandekar, V. M. 1955. Certain modified forms of binomial and Poisson distributions. Sankhya 15:237-50.

[9] Feller, W. 1968. An Introduction to Probability Theory and its Applications. Vol. I. Wiley, New York.

[10] Fisz, M. 1963. Probability Theory and Mathematical Statistics. Wiley, New York.

[11] Foster, F. G. 1952. A Markov chain derivation of discrete distributions. Ann. Math. Stat. 23:624-27.

[12] Haight, F. A. 1967. Handbook of the Poisson Distribution. Wiley, New York.

[13] Jones, P. C. T.; Mullison, J. E.; and Quenouille, M. H. 1947. A technique for the quantitative estimation of soil micro-organisms. J. Gen. Microbiol. 2:54-69.

[14] Karlin, S. 1966. A First Course in Stochastic Processes. Academic Press, New York.

[15] ________, and McGregor, J. 1958. Linear growth, birth and death processes. J. Math. Mech. 7:643-62.

[16] Kendall, D. G. 1948. On generalized birth and death processes. Ann. Math. Stat. 19:1-15.

[17] ________. 1949. Stochastic processes and population growth. J. Roy. Stat. Soc. Ser. B 11:230-64.

[18] Lundberg, O. 1964. On Random Processes and their Application to Sickness and Accident Statistics. Uppsala,

[19] Moran, P. A. P. 1968. An Introduction to Probability Theory. Clarendon Press, Oxford.

[20] Parzen, E. 1960. Modern Probability Theory and its Applications. Wiley, New York.

[21] Patil, G. P., and Joshi, S. W. 1968. A Dictionary and Bibliography of Discrete Distributions. Boyd, Edinburgh; Hofner Pub. Co., New York.

[22] Quenouille, M. H. 1949. A relation between the logarithmic, Poisson, and negative binomial series. Biometrics 5:162-64.

[23] Ramakrishnan, A. 1951. Some simple stochastic processes. J. Roy. Stat. Soc. Ser. B 13:131-40.

[24] Rogers, A. 1969. Quadrat analysis of urban dispersion. Working paper N. 93. Institute of Urban and Regional Development. U. California, Berkeley.

A PROBABILITY DISTRIBUTION ARISING IN A RIFF-SHUFFLE

V. R. RAO UPPULURI*
W. J. BLOT

Oak Ridge National Laboratory
Oak Ridge, Tennessee

SUMMARY

A discrete random variable $X = 0,1,2,\ldots,m-1$, with probability density function

$$\binom{m+x-1}{x} (p^m q^x + q^m p^x),\ p > 0,\ p+q = 1,$$

has arisen during the study of riff-shuffles as a Markov Process. An expression for the moment-generating function of this variable is obtained, from which the mean and variance are derived. By equating two different expressions for the mean of the variable, a closed form for the incomplete beta function with equal arguments is obtained as

$$I_p(m,m) = p + (p-q) \sum_{k=1}^{m-1} \binom{2k-1}{k} (pq)^k.$$

It is shown how this expression aids the rapid computation of the incomplete beta function with integral arguments. This expression is also used in deriving the asymptotic (m-large) expressions for the mean and variance. The standardized variate is shown to converge to the Gaussian distribution as $m \to \infty$. A result corresponding to the DeMoivre-Laplace type theorem is proved for m sufficiently large. Finally, applications are indicated to the genetic code problem, to Banach's matchbox problem, and to the World Series of baseball.

*Research sponsored by the U.S. Atomic Energy Commission under contract with the Union Carbide Corporation.

1. INTRODUCTION

The following discrete distribution has arisen while studying riff-shuffles as a Markov Process. Let us suppose that we have two decks of cards, designated deck A and deck B. Let each deck contain m cards. At each trial let us suppose that a card is picked with probability p from deck A, and with probability q from deck B, $p,q > 0$ and $p+q = 1$. We are interested in studying the random variable X which represents the number of cards chosen from one deck when all the m cards of the other deck have been chosen. In other words, the trials are stopped as soon as one of the decks is exhausted, and the number of cards left in the other deck is equal to m-X. The possible values of X are 0,1, 2,...,(m-1), since the experiment is terminated as soon as one of the decks is exhausted.

In the case of independent trials, from first principles we can deduce the probability density function of the random variable X to be

$$P[X=x] = \binom{m+x-1}{x}\left(p^m q^x + q^m p^x\right), \quad x = 0,1,2,\ldots,m-1. \tag{1}$$

The binomial coefficient corresponds to the fact that we are not concerned with the particular trial at which a given card from the unexhausted deck is chosen. The two terms in the parentheses correspond to the fact that we might exhaust either deck A or deck B, and these are two mutually exclusive events.

For the sake of completeness, in the next proposition we shall give an analytical proof of the fact that X is a genuine discrete random variable.

Proposition 1.1.

$$\sum_{x=0}^{m-1}\binom{m+x-1}{x}\left(p^m q^x + q^m p^x\right) = 1.$$

for $m \geq 1$, $p > 0$, $p+q = 1$.

Proof: We will first show that

$$\sum_{x=0}^{m-1}\binom{m+x-1}{x} p^m q^x = I_p(m,m) = \frac{1}{\beta(m,m)}\int_0^p t^{m-1}(1-t)^{m-1}dt, \tag{2}$$

where $\beta(m,m) = \dfrac{\Gamma(m)\Gamma(m)}{\Gamma(2m)}$;

and by symmetry we will have

$$\sum_{x=0}^{m-1}\binom{m+x-1}{x}q^m p^x = I_q(m,m), \tag{3}$$

which will prove the proposition, since $I_p(m,m)+I_q(m,m) = 1$. Consider the Taylor expansion about the origin (with remainder in the integral form): $f(q) = 1/(1-q)^r = 1/p^r$, $0<q<1$.

$$f(q) = f(0)+q[f'(0)]+\dots+\frac{q^{k-1}f^{(k-1)}(0)}{(k-1)} +$$

$$\frac{1}{(k-1)!}\int_0^q f^{(k)}(q-t)\, t^{k-1}\, dt$$

$$= 1+rq+\frac{r(r+1)}{2}q^2+\dots+\frac{r(r+1)\dots(r+k-q)}{(k-1)!}q^{k-1} +$$

$$\frac{1}{(k-1)!}\int_0^q \frac{r(r+1)\dots(r+k)}{(1-q+t)^{r+k}}t^{k-1}dt$$

$$= \sum_{x=0}^{k-1}\binom{r+x-1}{x}q^x+\frac{1}{\Gamma(k)}\int_0^q \frac{\Gamma(r+k)}{\Gamma(r)}\left(\frac{t}{1-q+t}\right)^{k-1}\left(\frac{1}{1-q+t}\right)^{r+1} dt$$

$$= \sum_{x=0}^{k-1}\binom{r+x-1}{x}q^x+\frac{1}{\beta(k,r)}\int_0^q \frac{u^{k-1}(1-u)^{r-1}}{(1-q)^r}dt,$$

where $u = t/(1-q+t)$,

$$= \sum_{x=0}^{k-1}\binom{r+x-1}{x}q^x+\frac{1}{p^r}I_q(k,r).$$

$$\sum_{x=0}^{k-1}\binom{r+x-1}{x}p^r q^x = 1-I_q(k,r) = I_p(r,k).$$

The special case when $r = k = m$ gives the required result.

This proposition shows that X is a discrete random variable with probability density function (p.d.f.) given by

$$f(x) = \binom{m+x-1}{x}\left(p^m q^x+q^m p^x\right),\quad x = 0,1,\dots,(m-1).$$

In the next section we shall discuss the moments of this variable.

2. MOMENT-GENERATING FUNCTION, EXPECTED VALUE AND VARIANCE

2.1 The Moment-Generating Function

Proposition 2.1.1: The moment-generating function of the variable X is given by

$$\varphi(t) = \frac{p}{1-qe^t}{}^m\left[1-I_{qe^t}(m,m)\right] + \left(\frac{q}{1-pe^t}\right)^m\left[1-I_{pe^t}(m,m)\right]. \tag{4}$$

Proof: (We shall only outline the proof of this proposition since the argument follows along the same lines as the proof of proposition 1.1.)

$$\varphi(t) = E\left[e^{tX}\right] = \sum_{x=0}^{m-1}\binom{m+x-1}{x}\left(p^m q^x + q^m p^x\right)e^{tx}$$

$$= \sum_{x=0}^{m-1}\binom{m+x-1}{x}p^m\left(qe^t\right)^x + \sum_{x=0}^{m-1}\binom{m+x-1}{x}q^m\left(pe^t\right)^x.$$

By considering the Taylor expansion of $f(q) = 1/(1-qe^t)^r$ about $q = 0$, it can be shown that

$$f(q) = \sum_{x=0}^{k-1}\binom{r+x-1}{x}\left(qe^t\right)^x + \frac{1}{(1-qe^t)^r}\, I_{qe^t}(k,r)$$

and it follows that

$$\sum_{x=0}^{m-1}\binom{m+x-1}{x}p^m\left(qe^t\right)^x = \left(\frac{p}{1-qe^t}\right)^m\left[1-I_{qe^t}(m,m)\right].$$

2.2. The Expected Value

We shall first derive the expected value of X by differentiating the moment-generating function and finding its value at the origin, and then derive it by direct computation. In this process we will show how we can obtain new identities involving the incomplete beta function.

Proposition 2.2.1.

$$\varphi'(0) = E(X) = \frac{m}{pq}\left[q^2 I_p(m,m) + p^2 I_q(m,m)\right] - \frac{1}{\beta(m,m)}(pq)^{m-1}. \tag{5}$$

Proof: This can be easily verified by direct computation of $\varphi'(t)$.

Proposition 2.2.2:

$$E(X) = \sum_{x=0}^{m-1} x\binom{m+x-1}{x}\left(p^m q^x + q^m p^x\right)$$

$$= \frac{m}{pq}\left[q^2 I_p(m+1,m+1)+p^2 I_q(m+1,m+1)\right]-\frac{4(pq)^m}{\beta(m,m)} \quad (6)$$

Proof:

$$\sum_{x=1}^{m-1} x\binom{m+x-1}{x}p^m q^x = p^m qm \sum_{\ell=0}^{m-2}\binom{m+\ell}{\ell}q^\ell$$

$$= p^m qm \sum_{\ell=0}^{m-2}\binom{m+1+\ell-1}{\ell}q^\ell$$

$$=p^m qm\left[\sum_{\ell=0}^{m}\binom{m+1+\ell-1}{\ell}q^\ell-\binom{2m}{n}q^m-\binom{2m-1}{m-1}q^{m-1}\right]$$

$$= p^m qm\left[\frac{I_p(m+1,m+1)}{p^{m+1}}-\frac{2m}{m^2}\cdot\frac{\Gamma(2m)}{\Gamma(m)\Gamma(m)}q^m-\frac{\Gamma(2m)}{m\Gamma(m)\Gamma(m)}q^{m-1}\right]$$

$$= \frac{mq}{p}I_p(m+1,m+1)-\frac{2}{\beta(m,m)}q^{m+1}p^m-\frac{1}{\beta(m,m)}q^m p^m .$$

Similarly we can get an expression for $\sum_{x=0}^{m-1} x\binom{m+x-1}{x}q^m p^x$, and adding these two we obtain the desired result.

2.3 A Procedure for Finding $I_p(a,b)$

By equating the two expressions for E(X), we will obtain a closed form for $I_p(m,m)$ as follows:

Proposition 2.3.1.

$$I_p(m+1,m+1) = p+(p-q)\sum_{k=1}^{m}\binom{2k-1}{k}(pq)^k,$$

for $p > 0$, $(p+q) = 1$. (7)

Proof:

$$E(X) = \frac{m}{pq}\left[q^2 I_p(m,m)+p^2 I_q(m,m)\right]-\frac{(pq)^{m-1}}{\beta(m,m)}, \text{ from (5)}$$

$$= \frac{m}{pq}\left[q^2 I_p(m+1,m+1)+p^2 I_q(m+1,m+1)\right]-\frac{4(pq)^m}{\beta(m,m)}, \text{ from (6)}$$

Therefore $[I_p(m+1,m+1)]-I_p(m,m)]\ (q^2-p^2) = \frac{(pq)^m}{m\,\beta(m,m)}(1-4pq)$,

and so

$$I_p(m+1,m+1) - I_p(m,m) = (p-q)(pq)^m / m\beta(m,m)$$

$$I_p(m,m) - I_p(m-1,m-1) = (p-q)(pq)^{m-1}/(m-1)\beta(m,m)$$

$$\vdots \qquad \vdots \qquad \vdots$$

$$I_p(2,2) - I_p(1,1) = (p-q)(pq).$$

Adding this system and noting $I_p(1,1) = p$, we have

$$I_p(m+1,m+1) = p+(p-q)\sum_{k=1}^{m} \frac{(pq)^k}{k\beta(k,k)}$$

$$= p+(p-q)\sum_{k=1}^{m} \binom{2k-1}{k}(pq)^k.$$

This expression for $I_p(m,m)$, namely

$$I_p(m,m) = p+(p-q)\sum_{k=1}^{m-1} \binom{2k-1}{k}(pq)^k, \tag{8}$$

seems to have been relatively unnoticed. The only closed form for $I_p(m,m)$, other than (2) and (8), appears in Pearson [8] and Abramowitz and Stegun [1] as follows:

$$I_p(m,m) = 1-\tfrac{1}{2}I_{4pq}(m,\tfrac{1}{2}), \text{ where } p+q = 1. \tag{9}$$

However, investigation reveals that this formula is valid only for $p \geq \frac{1}{2}$. Thus for the sake of completeness, we shall give this result as the next proposition.

Proposition 2.3.2.

$$I_p(m,m) = \tfrac{1}{2}I_{4pq}(m,\tfrac{1}{2}) \text{ for } \begin{cases} 0<p\leq 1/2 \\ p+q = 1 \end{cases}. \tag{10}$$

Proof: We note that

$$\beta(m,1/2) = \frac{\Gamma(m)\Gamma(1/2)}{\Gamma(m+1/2)} = \frac{2^m(m-1)!\Gamma(1/2)}{(2m-1)(2m-3)\ldots5(3)(1)\Gamma(1/2)}$$

$$= \frac{2^m(m-1)!2^{m-1}(m-1)!}{(2m-1)\ldots3(1)(2m-2)(2m-4)\ldots6(4)2} = \frac{4^m}{2}\beta(m,m).$$

And,

$$I_p(m,m) = \frac{1}{\beta(m,m)}\int_0^p x^{m-1}(1-x)^{m-1}dx$$

$$= \frac{4^m}{2\beta(m,1/2)}\int_0^p x^{m-1}(1-x)^{m-1}dx, \text{ from above,}$$

$$= \frac{1}{2}I_{4pq}(m,1/2).$$

The last step follows by the substitution $4x(1-x) = y$ and by noting that $x<p\leq\frac{1}{2}$.

From expression (7) and from two other recursive relations for the incomplete beta function, we will derive a procedure for computing $I_p(a,b)$, when a and b are positive integers and $0<p<1$. We need the following:

Proposition 2.3.3.

$$I_p(m,m)-I_p(m+k,m-k) = \sum_{j=0}^{k-1}\binom{2m-1}{m+j}p^{m+j}q^{m-1-j} \tag{11}$$

for $0<p<1$ and $p+q = 1$.

Proof: From the well-known relation between the incomplete beta function and cummulative binomial distribution (see, e.g., [7] page 264), we have

$$I_p(m,m) = \sum_{j=m}^{2m-1}\binom{2m-1}{j}p^jq^{2m-1-j},$$

$$I_p(m+1,m-1) = \sum_{j=m+1}^{2m-1} p^jq^{2m-1-j}, \text{ and}$$

$$I_p(m+2,m-2) = \sum_{j=m+2}^{2m-1}\binom{2m-1}{j}p^jq^{2m-1-j} .$$

Form the differences:

$$I_p(m,m) - I_p(m+1,m-1) = \binom{2m-1}{m}p^mq^{m-1}$$

$$I_p(m+1,m-1) - I_p(m+2,m-2) = \binom{2m-1}{m+1}p^{m+1}q^{m-2}$$

$$\vdots \qquad \vdots \qquad \vdots$$

Adding k such equations we get

$$I_p(m,m)-I_p(m+k,m-k) = \sum_{j=m}^{m+k-1} \binom{2m-1}{j} p^j q^{2m-1-j}$$

$$= \sum_{j=0}^{k-1} \binom{2m-1}{m+j} p^{m+j} q^{m-1-j} .$$

We also need the well-known result

$$I_p(a,b) = pI_p(a-1,b)+qI_p(a,b-1). \tag{12}$$

We shall now describe the procedure for calculating $I_p(a,b)$, where a and b are positive integers. We will have two cases, determined by whether a+b is even or odd. Without loss of generality we will assume that a > b; otherwise we will have to consider $1-I_q(b,a)$.

Case 1: a+b is even. Calculate $I_p(m,m)$ where $m = (a+b)/2$ using relation (8). Then using (11) compute $I_p(m+k,m-k)$ where $k = (a-b)/2$. This will give us the desired value of $I_p(a,b)$ since $m+k = a$ and $m-k = b$. Thus for example, to find $I_p(7,3)$ compute $I_p(5,5)$ using (8), and then find $I_p(7,3)$ using (11). This procedure is illustrated graphically in Figure 1(a).

Case 2: a+b is odd. Calculate $I_p(m,m)$ using (8) where $m = (a+b-1)/2$, then calculate $I_p(m+k-1,m-k+1)$ and $I_p(m+k,m-k)$, $k = (a-b+1)/2$, using (11). These values together with relation (12) give us $I_p(a,b)$. Thus, to find $I_p(7,2)$, for example, first find $I_p(4,4)$ using (8); then compute $I_p(6,2)$ and $I_p(7,1)$ using (11). Using (12), we obtain $I_p(7,2) = pI_p(6,2)+qI_p(7,1)$. This procedure is also illustrated graphically in Figure 1(b).

This procedure of finding $I_p(a,b)$ seems to be suitable for high-speed digital computers. We shall give a computer program (FORTRAN) in the appendix. We would also like to note the availability of [8] for finding $I_p(a,b)$ when $a,b \leq 50$ and $p = .01(.01).99$.

2.4 The Variance

From the moment-generating function (4), the variance of the distribution may be obtained. Equation (4) states that

$$\varphi(t) = \left(\frac{p}{1-qe^t}\right)^m \left[1-I_{qe^t}(m,m)\right] + \left(\frac{q}{1-pe^t}\right)^m \left[1-I_{pe^t}(m,m)\right] .$$

Equation (5) states that

$$\varphi'(0) = \frac{m}{pq}\left[q^2 I_p(m,m)+p^2 I_q(m,m)\right] - \frac{1}{\beta(m,m)}(pq)^{m-1} .$$

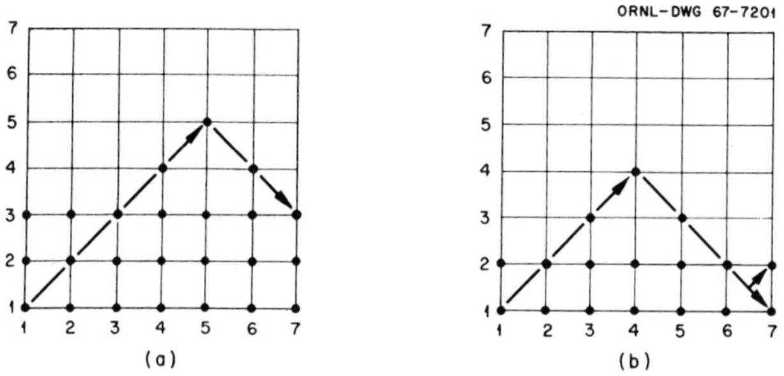

Fig. 1. (a) Demonstrates the procedure for finding $I_p(7,3)$. Six steps and one initial value are required (where 14 steps and 7 initial values were needed formerly). Geometrically what the procedure does is to move up the diagonal of the a x b array to the point (5,5), then proceed downwards at a 90° angle, 2 steps to $I_p(7,3)$. This diagonal travel allows faster computation than does stepwise movement. (b) Demonstrates the procedure for finding $I_p(7,2)$. Seven steps and one initial value are required. (Formerly 7 steps and 7 initial values were required).

Since $\text{Var}(X) = \varphi''(0)-[\varphi'(0)]^2$, we need only find $\varphi''(0)$, given in the following proposition, which may be verified by direct computation.

Proposition 2.4.1.

$$\varphi''(0) = \frac{m}{pq}\left[q^2 I_p(m,m) + p^2 I_q(m,m)\right] + \frac{m(m+1)}{p^2p^2}\left[q^4 I_p(m,m)+p^4 I_q(m,m)\right] - \frac{m}{\beta(m,m)}(pq)^{m-2}(1-2pq)-\frac{1}{\beta(m,m)}(pq)^{m-2}(1-3pq). \qquad (13)$$

The variance of X may also be obtained by computing $E[X(X-1)]$ since

$$\text{Var}(X) = E[X(X-1)]+E(X)-[E(X)]^2.$$

Proposition 2.4.2.

$$E[X(X-1)] = \frac{m(m+1)}{p^2q^2}\; q^4 I_p(m+2,m+2)+p^4 I_q(m+2,m+2) - \frac{(pq)^m}{\beta(m,m)}[16m(1-2pq)+4(1-4pq)]. \qquad (14)$$

Proof: The proof is similar to that of proposition 2.2.2.

A comparison of the two forms of variance may give further identities involving the incomplete beta function.

3. ASYMPTOTIC EXPRESSIONS FOR E(X) AND VAR(X)

3.1 Asymptotic Expression for $I_p(m,m)$

The probability density function f(x) given by (1) depends on the parameters m and $0 < p < 1$. In this section we shall derive the asymptotic (when m is large) expressions for the mean and variance. Before that, though, we shall prove the following:

Lemma 3.1.1.

$$\lim_{m\to\infty} I_p(m,m) = \begin{cases} 0 & p < 1/2 \\ 1/2 \text{ for } & p = 1/2 \\ 1 & p > 1/2 \end{cases} . \tag{15}$$

Proof: First we will note, for $p \neq \frac{1}{2}$.

$$\sum_{k=0}^{\infty} \binom{2k}{k} (pq)^k = \sum_{k=0}^{\infty} \binom{1/2 + k-1}{k} (4pq)^k$$

$$= \frac{1}{\sqrt{1-4pq}} = \frac{1}{|p-q|} .$$

From (8) we have

$$\lim_{m\to\infty} I_p(m,m) = p+(p-q) \sum_{k=1}^{\infty} \binom{2k-1}{k} (pq)^k$$

$$= p+\frac{1}{2}(p-q)\left[\sum_{k=0}^{\infty} \binom{2k}{k} (pq)^k - 1\right]$$

$$= p+\frac{1}{2}(p-q)\left[\frac{1}{p-q} - 1\right] \text{if } p > q$$

$$= p+\frac{1}{2}(p-q)\left[\frac{1}{q-p} - 1\right] \text{if } q > p$$

$$= \begin{cases} 1 \text{ if } p > 1/2 \\ 0 \text{ if } p < 1/2 . \end{cases}$$

Since $I_p(m,m) = 1/2$ for all m, when $p = 1/2$, we will have $\lim_{m\to\infty} I_p(m,m) = 1/2$ when $p = 1/2$.

This lemma, together with (9) and (10), implies that

$\lim_{m\to\infty} I_{4pq}(m,\frac{1}{2}) = 0$, where $\frac{1}{2} > p > 0$ and $p+q = 1$.

3.2 The Asymptotic Mean

Next we derive the asymptotic expression for the mean of the distribution. For the remainder of this paper we will assume, without loss of generality, that $p \geq q$ (equivalent to $p \geq 1/2$), since f(x) given in (1) remains unchanged when p and q are interchanged (a demonstration of its symmetry). Hence the results when $p < q$ are identical to the results when $p > q$ with p and q interchanged. We will use the notation $f \approx g$, whenever $\lim_{m\to\infty}(f/g) = 1$.

Proposition 3.2.1.

For large m,

$$E(X) \approx \begin{cases} m-2\sqrt{m/\pi} & \text{when } p = \frac{1}{2} = q \\ mq/p & \text{when } p > \frac{1}{2} \end{cases} . \qquad (16)$$

Proof: First consider the case $p = q = 1/2$. From (5) we have

$$E(X) = m-\frac{1}{(m,m)}\left(\frac{1}{2^{2m-2}}\right) .$$

By using Stirling's approximation for factorial, one can show that

$$\frac{1}{\beta(m,m)} \approx \frac{2^{2m-2}(2m-1)}{\sqrt{(m-1)\pi}} . \qquad (17)$$

Therefore, we have for large m, when $p = 1/2$,

$$E(X) \approx m-2\sqrt{m/\pi} .$$

In the general case, from (5) we have

$$E(X) = \frac{m}{pq}\left[q^2 I_p(m,m)+p^2 I_q(m,m)\right] - \frac{1}{\beta(m,m)}(pq)^{m-1}$$

$$\approx \frac{m}{pq}[q^2 I_p(m,m)+p^2 I_q(m,m)]-\frac{(2m-1)}{\sqrt{(m-1)\pi}}(4pq)^{m-1} \text{ from (17)},$$

$$\approx \frac{m}{pq}q^2 = \frac{mq}{p},$$

since from lemma (3.1.1), $\lim_{m\to\infty} I_q(m,m) = 0$, and in the second term $(4pq)^{m-1} = e^{(m-1)\log(4pq)}$, with $\log(4pq) < 0$, has the dominating effect.

3.3. The Asymptotic Variance

In this section, we will derive the asymptotic expression for the variance of the distribution.

Proposition 3.3.1. For large m,

$$\mathrm{Var}(X) = \begin{cases} 2m - \frac{4m}{\pi} - 2\sqrt{m/\pi} & \text{for } p = 1/2 \\ \\ mq/p^2 & \text{for } m > 1/2 \end{cases} \tag{18}$$

Proof: Let us consider the case p = q = 1/2. Using (13) we get

$$\mathrm{Var}(X) = 2m - \frac{1}{\beta(m,m)}\left(\frac{1}{2^{2m-2}}\right) - \frac{1}{\beta^2(m,m)}\left(\frac{1}{2^{4m-4}}\right)$$

$$\approx 2m - 2\sqrt{m/\pi} - 4m/\pi$$

since from (17), $\frac{1}{\beta(m,m)}\left(\frac{1}{2^{2m-2}}\right) \approx 2\sqrt{m/\pi}$.

In the general case we have from (13) and (5),

$$\varphi''(0) = \frac{m}{pq}\left[q^2 I_p(m,m) + p^2 I_q(m,m)\right] +$$

$$\frac{m(m+1)}{p^2q^2}\left[p^4 I_p(m,m) + q^4 I_p(m,m)\right] -$$

$$\frac{m}{\beta(m,m)}(pq)^{m-2}(1-2pq) - \frac{(pq)^{m-2}}{\beta(m,m)}(1-3pq),$$

and

$$[\varphi'(0)]^2 = \frac{m^2}{(pq)^2}\left[q^4 I_p^2 + 2p^2q^2\left(I_p I_q\right) + p^4 I_q^2\right] -$$

$$2m\left[q^2 I_p + p^2 I_q\right]\frac{(pq)^{m-2}}{\beta(m,m)} + \frac{(pq)^{2m-2}}{\beta^2(m,m)},$$

and $\mathrm{Var}(X) = \varphi''(0) - [\varphi'(0)]^2$.

By using lemma (3.1.1), together with (17), and noting that $4pq < 1$, it can be shown that

$$\mathrm{Var}(X) \approx mq/p^2 \quad \text{if } p > q.$$

3.4 A Comparison of Exact and Asymptotic Expressions

In Table 1 we present the exact values of the mean, E(X), and variance, V(X), and their asymptotic values (mq/p) and (mq/p^2) respectively, for p = .5(.1).9 and m = 20, 50, 100. It is interesting to note that the asymptotic (m-large) values agree more closely with the exact values when p is close to 1 (and equivalently when p is close to 0). For example, the asymptotic values seem to be quite precise for p = 0.9 even when m is as small as 10.

TABLE 1. Comparison of the Exact and Asymptotic Values of the Expected Value E(X) and Variance V(X)

p	m	E(x)	mq/p	V(x)	mq/p^2
0.5*	20	14.9852	14.9550	9.8367	8.4890
	50	42.0411	42.0210	28.6966	28.3600
	100	88.7303	88.7160	61.7242	61.3920
0.6	20	12.7254	13.3333	14.4219	22.2222
	50	33.1703	33.3333	50.3723	55.5556
	100	66.6486	66.6667	109.9467	111.1111
0.7	10	4.1720	4.2857	4.9841	6.1225
	20	8.5540	8.5714	11.8811	12.2445
	50	21.4285	21.4286	30.6084	30.6122
	100	42.8571	42.8571	61.2245	61.2245
0.8	10	2.4957	2.5000	3.0696	3.1250
	20	5.0000	5.0000	6.2489	6.2500
	50	12.5000	12.5000	15.6250	15.6250
	100	25.0000	25.0000	31.2500	31.2500
0.9	10	1.1111	1.1111	1.2344	1.2346
	20	2.2222	2.2222	2.4691	2.4691
	50	5.5556	5.5556	6.1728	6.1728
	100	11.1111	11.1111	12.3457	12.3457

*Asymptotic formulae for p = 0.5 are given by $E(X) \approx m-2\sqrt{m/\pi}$ and $V(X) \approx 2m-2\sqrt{m/\pi}-4m/\pi$.

4. LIMIT THEOREMS

4.1 The Asymptotic (m-large) Distribution of the Standardized Variate

Having obtained expressions for the asymptotic mean and variance of the random variable X with probability density function f(x) given by (1), we will now study the limiting behavior of the standardized variate $Z_m = [X-E(X)]/\sqrt{Var(X)}$, as $m \to \infty$. In the next proposition we will show that Z_m is asymptotically equivalent in distribution to Y_m, which is obtained from Z_m by replacing E(X) and Var(X) by their asymptotic expressions, obtained in propositions (3.2.1) and (3.3.1); i.e.,

$$Y_m = [X-mq/p]/\sqrt{mq/p^2} .$$

Proposition 4.1.1.

Y_m and Z_m have the same limiting distribution as $m \to \infty$.

Proof: $Z_m = [X - \frac{mq}{p} + \frac{mq}{p} - E(X)]/\sqrt{Var(X)}$.

$$a_m Z_m = Y_m + b_m,$$

where $a_m = \sqrt{Var(X)}/(\sqrt{qm/p})$ and

$$b_m = [(qm/p) - E(X)]/(\sqrt{qm/p}).$$

From proposition (3.2.1) we have that $a_m \to 1$ as $m \to \infty$. We will now show that $b_m \to 0$ as $m \to \infty$.

From (5), (15), and (17) we get

$$b_m = \frac{p}{\sqrt{qm}}\left[\frac{(4pq)^{m-1}(2m-1)}{\sqrt{\pi(m-1)}}\right]$$

$$\approx \frac{p}{\sqrt{q\pi}}(4pq)^{m-1} \to 0 \text{ as } m \to \infty, \text{ since } 4pq < 1.$$

By using a weaker version of Slutsky's theorem (see, e.g., [3], page 234) we conclude that Z_m and Y_m have the same limiting distribution.

In the next proposition we prove the asymptotic normality of Y_m, which, in turn, will imply the asymptotic normality of Z_m.

Proposition 4.1.2.

$$\lim_{m\to\infty} P[Y_m \le y] = \frac{1}{\sqrt{2\pi}}\int_{-\infty}^{y} e^{-t^2/2}dt,$$

where $Y_m = [X-(mq/p)]/(\sqrt{qm/p})$.

Proof: We will show that the sequence of moment-generating functions $\psi_m(t) = E[e^{tY_m}]$ (which exist since Y_m takes on only a finite set of values), converges as $m \to \infty$ to the moment-generating function of the standard Gaussian variable. Then by applying the central limit theorem in terms of moment-generating functions (see [6], p. 193) we have our proposition.

From proposition (2.1.1) the moment-generating function of X is given by

$$\varphi(t) = E\left[e^{tX}\right] = \left(\frac{p}{1-qe^t}\right)^m \left[1-I_{qe^t}(m,m)\right] + \left(\frac{q}{1-pe^t}\right)^m \left[1-I_{pe^t}(m,m)\right].$$

By changing variables and using Y_m, where

$$Y_m = (pX-mq)/\sqrt{qm},$$

we will have

$$\psi_m(t) = e^{-t\sqrt{qm}}\,\varphi(\Theta), \qquad \text{where } \Theta = pt/\sqrt{qm},$$

$$= e^{-t\sqrt{qm}}(a_m)\left(1+\frac{b_m}{a_m}\right),$$

where $$a_m = \left(\frac{p}{1-qe^{\Theta}}\right)\left(1-I_{qe^{\Theta}}\right),$$

and $$b_m = \left(\frac{q}{1-pe^{\Theta}}\right)\left(1-I_{pe^{\Theta}}\right).$$

We note that for $p > 1/2$, $b_m < a_m$, and also for large m, $b_m \approx O(a_m)$. Therefore $\log \psi_m(t) = -t\sqrt{qm}+\log a_m+\log(1+\frac{b_m}{a_m})$.

It can be shown that

$$\log a_m = t\sqrt{qm}+t^2/2+\log\left[1-I_{qe^{pt/\sqrt{qm}}}\right].$$

And therefore $\lim_{m\to\infty}\psi(t) = e^{t^2/2}$, which shows that

$$\lim_{m\to\infty} P[Y_m \leq y] = \frac{1}{\sqrt{2\pi}}\int_{-\infty}^{y} e^{-x^2/2}dx.$$

4.2 Graphic Analysis

In this section we will give graphs of f(x) plotted against x for the case m = 51 and p = 0.5, 0.6, 0.7 in Figure 2. It is interesting to note that the probability density function f(x), given by (1), has an extreme shape when p = 0.5, becomes almost symmetrical when $0.6<p<0.7$, and changes shape again as p approaches 1. Figures 3(a) and 3(b) show the graphs of f(x) when p = 2/3 and m = 51 and m = 81, respectively. We note that almost symmetric nature of f(x) about x = m/2. This suggests the possibility of proving a DeMoivre-Laplace type limit theorem for f(x) (see [4], page 168). In the next section we will derive an asymptotic (m-large) expression for the probability density function f(x).

4.3 DeMoivre-Laplace Type Limit Theorem

Proposition 4.3.1.

$$f(x) = \binom{m+x-1}{x}(p^m q^x + q^m p^x) \approx g\left(\frac{x-mq/p}{\sqrt{mq/p}}\right),$$

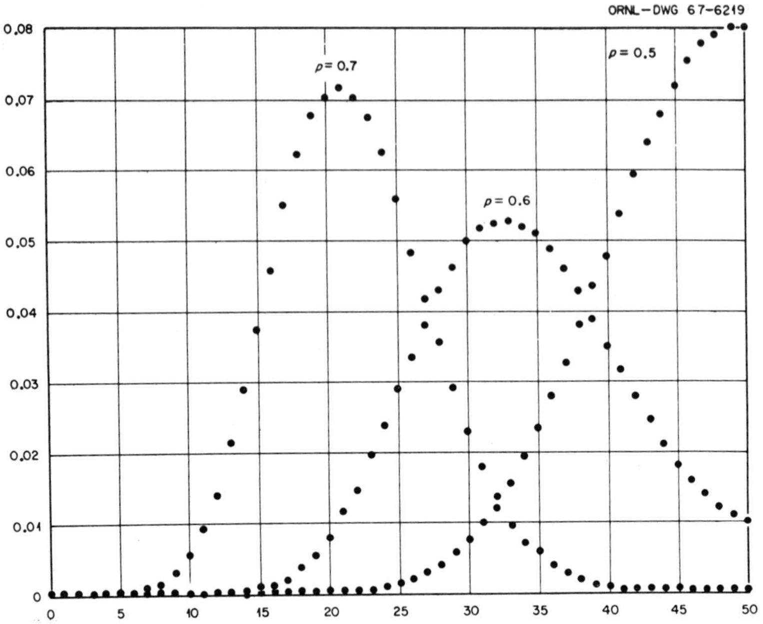

Fig. 2. $Pr[X=x] = \binom{m+x-1}{x}(p^m q^x + q^m p^x)$, $x = 0,1,2,\ldots,m-1$, when $m = 51$ for $p = 0.5$, 0.6, and 0.7.

where $g(u) = \frac{1}{\sqrt{2\pi}}e^{-u^2/2}$, when $m \to \infty$ and $x \to \infty$ in such a way that

$$\frac{(x-mq/p)^3}{m^2(\sqrt{q/p})^3} \to 0.$$

Proof: Using Stirling's formula for factorials, we have

$$f(x) \approx (p^m q^x + q^m p^x)\left[\frac{m-1}{2\pi x(m+x-1)}\right]^{1/2}\left(\frac{m+x-1}{m-1}\right)^m \left(\frac{m+x-1}{x}\right)^m .$$

Letting $\delta = \delta_x = x-mr$, where $r = q/p$, and rearranging we obtain

$$f(x) \approx r^m(r^{-m}+r^{-mr-\delta})\left[\frac{m-1}{2\pi(\delta+mr)(m+mr+\delta-1)}\right]^{1/2} \text{ mult. by}$$

$$\left[\frac{m+mr+\delta-1}{m(r+1)}\Big/\frac{r(m-1)}{m(r+1)q}\right]^m \left[\frac{m+mr+\delta-1}{(mr+q)/\delta}\right]^{mr+\delta}$$

Considering the last two terms, and calling their product A, we have

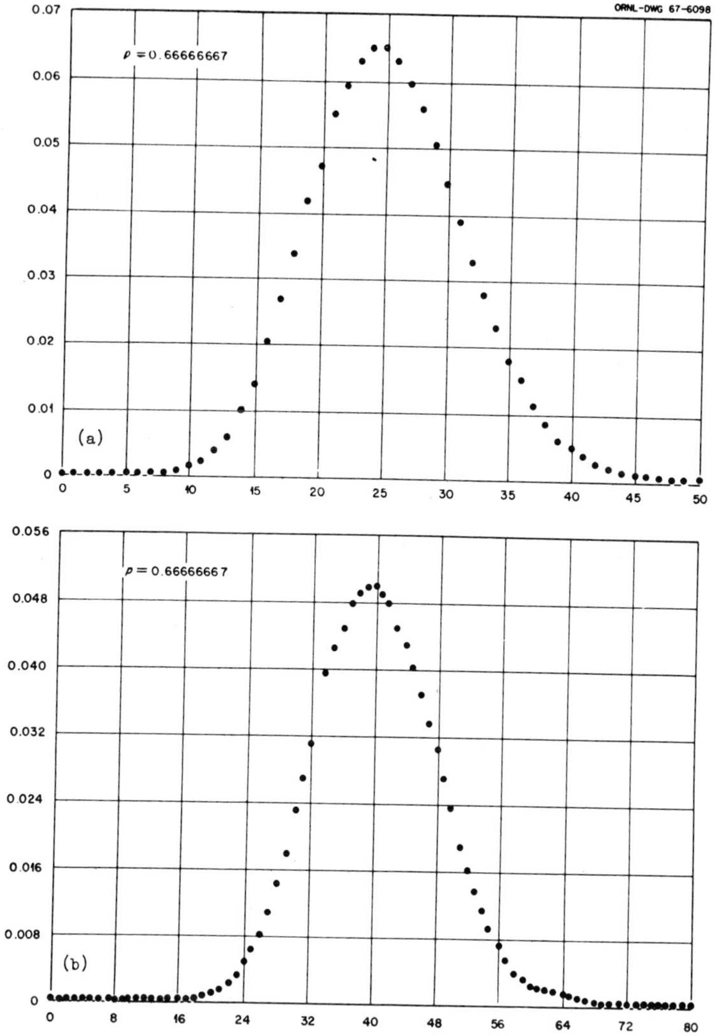

Fig. 3. (a) $\Pr[X=x] = \binom{m+x-1}{x}\left(p^m q^x + q^m p^x\right)$ when m = 51 for p = 2/3. (b) $\Pr[X=x] = \binom{m+x-1}{x}\left(p^m q^x + q^m p^x\right)$ when m = 81 for p = 2/3.

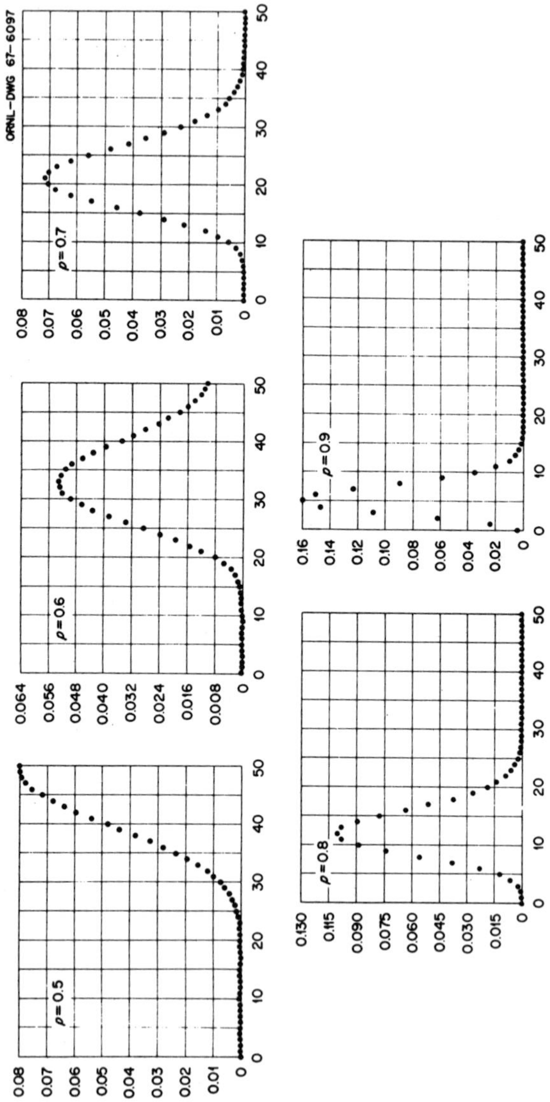

Fig. 4. Pr[X=x], p = 0.5, p = 0.6, p = 0.7, p = 0.8, and p = 0.9.

$$A = \left[\frac{1+\frac{\delta-1}{m(r+1)}}{1-1/m}\right]^{m} \left[\frac{1+\frac{\delta-1}{m(r+1)}}{1+\delta/mr}\right]^{mr+\delta}$$

and so,

$$\log A = (m+mr+\delta)\log\left[1-\frac{1-\delta}{m(r+1)}\right]-m\log(1-1/m) -$$

$$(mr+D)\log(1+\delta/mr).$$

Expanding the logarithmic functions in series, and using the assumption

$$\frac{(x-mq/p)^3}{m^2(\sqrt{q/p})^3} \approx 0$$

for large m, we obtain

$$\log A \approx -p^2\delta^2/2mq,$$

and hence, for large m

$$A \approx \exp -\left[\frac{1}{2}\,\frac{(x-mq/p)^2}{mq/p^2}\right].$$

Considering the first two terms, and calling their product B, we have

$$B = r^m(r^{-m}+r^{-mr-\delta})\left[\frac{m-1}{2\pi(mr+\delta)(m+mr+\delta-1)}\right]^{1/2}.$$

Now, for large m, $mr+\delta \approx mr$.

Hence, for large m,

$$B \approx \left[1+r^{m(1-r)}\right]\left(\frac{1}{2\pi mq/p^2}\right)^{1/2},$$

and therefore

$$f(x) = AB \approx \frac{1}{\sqrt{2\pi mq/p^2}}\exp -\left[\frac{1}{2}\,\frac{(x-mq/p)^2}{mq/p^2}\right].$$

5. APPLICATIONS

5.1. Baseball

One conceivable application of this probability distribution could be to the World Series between the American League and the National League baseball champions. It may be of interest to know the total number of games played in a particular World Series (see [9], p. 93, problem 2). We shall identify the parameters in our distribution as follows:

m = 4 = the number of games necessary by a team to win the series.

p = Probability for the American League to win a game in the series.

q = Probability for the National League to win a game in the series.

The series will come to an end as soon as one of the teams win four games; by that time the other team would have won three or fewer games, this quantity comprising our random variable X. In other words, 4+X will be the number of games actually played in a series, where X can take the values 0,1,2, and 3.

The probability that the American League wins the series while losing only x games is

$$\binom{3+x}{x}p^4q^x, \quad x = 0,1,2,3;$$

and the probability that the American League loses the series while winning only x games is

$$\binom{3+x}{x}q^4p^x, \quad x = 0,1,2,3.$$

Hence

$$f(x) = \binom{3+x}{x}\left(p^4q^x+q^4p^x\right), \quad x = 0,1,2,3,$$

gives the probability that the series goes exactly 4+x games. We note, incidentally, that f(x) gives the probability without specifying which team wins. The probability that the American League wins the series is given by

$$\sum_{x=0}^{3}\binom{3+x}{x}p^4q^x$$

and the probability that the American League loses the series is given by

$$\sum_{x=0}^{3}\binom{3+x}{x}q^4p^x.$$

Finally we find that

Prob. [American League wins the series] +
Prob. [American League loses the series]

$$= \sum_{x=0}^{3}\binom{3+x}{x}\left(p^4q^x+q^4p^x\right) = 1.$$

We have the corresponding probabilities for the National League.

Since 1905, excluding the series during the years 1919-1921 when five wins were required to win, the American League has won 36 World Series while losing 23 up to the time of this writing. In the 59 series, the American League has won 188 games and has been defeated in 148 games. So an estimate of p is

$$\hat{p}_1 = \frac{188}{336} = .56;$$

using this estimate of p, the duration of the coming World Series can be predicted. The following values of f(x), x = 0,1,2,3, give the probabilities that this coming World Series will last precisely 4 + x games: f(0) = .136; f(1) = .257; f(2) = .308; f(3) = .299.

Before the start of each series, Las Vegas odds, based on the competing teams' performances during the season, giving the probability that either the American League team or National League team will win the series are stated. That is, π is given such that

$$\sum_{x=0}^{3} \binom{3+x}{x} p^4 q^x = \pi.$$

We have shown that $\sum_{x=0}^{3} \binom{3+x}{x} p^4 q^x = I_p(4,4) = \pi.$

Thus from tables of percentage points of the incomplete beta function we can find p. Then using this value of p, call it $\hat{p}_2$, the duration of the World Series can be predicted. For example, consider the 1965 series where Los Angeles was quoted at 2-3 to win (see, e.g., Rothman [9]). Finally, π = 0.6, and from the tables [2] $\hat{p}_2$ is seen to be 0.5461. Therefore f(0) = 0.131, f(1) = 0.254, f(2) = 0.310, and f(3) = 0.305.

5.2 Banach Match Box Problem

In a slightly different form the probability density function f(x) given by (1) has been considered in [4], for the case p = 1/2, referred to as the "Banach Match Box Problem." In the problem, let u_r denote the probability that r matches remain in one pocket when the other pocket is discovered empty, given that we have N matches initially in each pocket and that the pockets are chosed with equal probability. On page 157, [4] shows that

$$u_r = \binom{2N-r}{N} (1/2)^{2N-r}, \quad r = 0,1,2,\ldots,N.$$

This corresponds to f(r) in (1) with the following identification of the variable and the parameters:

$$N = m-1, \; r = m-1-x, \text{ and } p = q = 1/2.$$

On page 213, [4] shows that

$$\sum_{r=0}^{N} ru_r = \frac{2N+1}{2^{2N}}\binom{2N}{N}-1,$$

which can also be deduced from (5) by the above identification of parameters. Feller [4] also gives an expression for the asymptotic mean as $2\sqrt{\frac{N}{\pi}}-1$, which can be deduced from (16).

5.3 The Genetic Code Problem

The location of the four nucleotides denoted by G, T, A, and C, on the DNA strand of any individual seem to determine uniquely the genetic characteristics of that individual. If we consider the genetic progeny of two mated individuals to be a riff-shuffle between the two strands, it may be of interest to know the number of nucleotides of one of the strands left out at the end--without affecting the order of the nucleotides because of the riff-shuffle. This might indicate a kind of dominance. This is an oversimplified way of looking at the very important problem of the genetic code, and perhaps one has to gain more insight into the following problems: (1) the number of nucleotides on a DNA strand, (2) whether their cardinality is the same within the same species, (3) whether one can look at the DNA strand of the progeny and identify the dominating parts of the parents, and (4) how many generations such genetic characteristics can persist. The answers may well lead to some solutions to problems associated with the genetic code; or on the other hand, they might show that mating is not a valid example of a riff-shuffle operation, after all.

ACKNOWLEDGMENTS

The authors would like to express their thanks to Mr. P. M. Kannan for plotting the probability density function using the calcomp computer; and to Professor L. R. Shenton and Mr. John R. Doner for some helpful discussions in the early part of this project; and to Dr. M. A. Kastenbaum about possible applications.

REFERENCES

[1] Abramowitz, M., and Stegun, I. 1964. Handbook of Mathematical Functions. National Bureau of Standards Applied Mathematics, Series 55. Department of Commerce, Washington.

[2] Bracken, Jerome 1966. Percentage Points of the Beta Distribution for use in Bayesian Analysis of Bernoulli Processes. Technometrics, 8:687-694.

[3] Cramer, Harald 1945. Mathematical Methods of Statistics. Princeton University Press, Princeton.

[4] Feller, William 1957. An Introduction to Probability Theory and its Applications, Vol. 1 (2nd ed.). Wiley, New York.

[5] Freund, John 1962. Mathematical Statistics. Prentice Hall, Englewood Cliffs, N. J.

[6] Hogg, R. V. and Craig, A. T. 1965. Introduction to Mathematical Statistics. (2nd ed.) Macmillan, New York.

[7] Owen, D. B. 1962. Handbook of Statistical Tables. Addison Wesley, Reading, Mass.

[8] Pearson, Karl 1934. Tables of the Incomplete Beta Function. Biometrika Office, London.

[9] Rothman, David 1967. "The Principle of Significant Inconsistency: How to Profit from Dependent Events." American Statistician, 21:23-24.

APPENDIX

A FORTRAN Program for the Computation of $I_p(a,b)$

```
      PROGRAM BETA
      DIMENSION F(200)
      F(1)=0.
   15 READ (50,1)M1,M2,P
    1 FORMAT(2I3,F3.2)
      M3=2*M1
      DO 10 K=2,M3
      XK=K
   10 F(K)=F(K-1)+LOGF(XK)
      Q=1.-P
      PQ=P*Q
      G=0.
      Y1=0.
      L=(M1+M2)/2
      IF(L-(M1+M2)/2.)20,21,21
   21 L1=L
      L2=(M1-M2)/2
      GO TO 22
   20 L1=(M1+M2-1)/2
      L2=(M1-M2+1)/2
   22 L11=L1-1
      DO 11 K=2,L11
      X1=F(2*K-1)-F(K)-F(K-1)+K*LOGF(PQ)
   11 Y1=Y1+EXPF(X1)
      BETAMM=P+(P-Q)*(Y1+P*Q)
      IF(L2)30,41,30
   30 DO 12 K=1,L2
      G1=F(2*L1-1)-F(L1+K-1)-F(L1-K)+(L1+K-1)*LOGF(P)+(L1-K)*LOGF(Q)
   12 G=G+EXPF(G1)
      IF(L-(M1+M2)/2.)40,41,41
   41 BETAAB=BETAMM-G
      GO TO 50
   40 BETAAB=P*(BETAMM-G+EXPF(G1))+Q*(BETAMM-G)
   50 WRITE(51,2)M1,M2,BETAAB
    2 FORMAT(2I5,E20.8)
      END
```

A METHOD FOR FINDING JOINT MOMENTS WHICH IS APPLICABLE TO A CLASS OF MULTIVARIATE DISCRETE DISTRIBUTIONS ON A SIMPLEX

BRUCE HOADLEY

Bell Telephone Laboratories, Inc.
Holmdel, New Jersey

SUMMARY

It is shown that Tsao's (1965) method for finding moments of a multivariate hypergeometric distribution is applicable to a large class of multivariate discrete distributions. The method is then applied to the compound multinomial distribution which arose (Hoadley (1969)) as both a prior and posterior distribution in a Bayesian approach to a categorical data problem.

1. INTRODUCTION

To calculate the moments of a multivariate hypergeometric distribution, Tsao [1965] introduces a special kind of generating function, from which the moments may be obtained by several differentiations. It is shown in this paper that his method is general and may be applied to any multivariate discrete distribution which can be characterized as the conditional distribution of a vector of independent discrete random variables $(Z_1,\ldots,Z_r)$, given $\sum_j Z_j = m$. Included in this class are the multivariate hypergeometric, multinomial, and compound multinomial distributions. In Section 3 the method is applied to the compound multinomial distribution to yield moments up to the 10th order. The need for these moments arose in a Bayesian approach to some categorical data problems (Hoadley [1969]) where both the prior and posterior distributions are compound multinomial.

2. THE GENERAL METHOD

Let $Z_1,\ldots,Z_r$ be independent discrete random variables, assuming values on the non-negative integers. Let

$$p_j(k) = P\{Z_j=k\} \text{ and } \sum_{k=0}^{M_j} p_j(k) = 1$$

(M_j could be $+\infty$). For $m \leq M*$ $(M* = \sum_j M_j)$, let

$$p(k_1,\ldots,k_r) = P\{(Z_1,\ldots,Z_r) = (k_1,\ldots,k_r) | Z* = m\}$$
$$= \left[\prod_j p_j(k_j)\right] \Big/ P\{Z* = m\}. \tag{2.1}$$

This defines a distribution on

$$S(m) = \underset{\sim}{k} = (k_1,\ldots,k_r): \ k_j \text{ is an integer,}$$
$$0 \leq k_j \leq M_j, \text{ and } k* = m. \tag{2.2}$$

Thoroughout the paper, $\underset{\sim}{Y}$ will denote a random vector whose distribution is defined by (2.1) and (2.2). Table 1 gives some common examples.

Now let $g_j(w)$ be the p.g.f. of Z_j and define for $\left|ue^{t_j}\right| < 1$

$$G(\underset{\sim}{t};u) = \prod_j g_j\left(ue^{t_j}\right). \tag{2.3}$$

This function can be viewed as a function which generates the moment-generating function for $p(\underset{\sim}{k})$ as is shown by Theorem 1.

TABLE 1. Examples of Formula (2.1).

Name	Name	Symbolic Name for $p(k_1,\dots,k_r)$
$p_j(k)$	$p(k_1,\dots,k_r)$	
Poisson	Multinomial	
$\dfrac{\lambda_j^k e^{-\lambda_j}}{k!}$	$m! \prod_j \dfrac{(\lambda_j/\lambda_*)^{k_j}}{k_j!}$	$\mathrm{Mtn}(m;\underset{\sim}{\lambda}/\lambda_*)$
Binomial	Multivariate Hypergeometric	
$\binom{M_j}{k} q^k (1-q)^{M_j-k}$	$\prod_j \binom{M_j}{k_j} \Big/ \binom{M_*}{m}$	$\mathrm{MH}(m;\underset{\sim}{M})$
Compound Poisson or Negative Binomial	Compound Multinomial	
$\dfrac{\Gamma(k+a_j)}{k!\Gamma(a_j)} q^k (1-q)^{a_j}$	$\dfrac{m!\Gamma(a_*)}{\Gamma(m+a_*)} \prod_j \dfrac{\Gamma(k_j+a_j)}{k_j!\Gamma(a_j)}$	$\mathrm{CMtn}(m;\underset{\sim}{a})$

Theorem 1. Let $M(\underset{\sim}{t};m)$ be the m.g.f. of $p(\underset{\sim}{k})$. Then

$$M(\underset{\sim}{t};m) = \frac{1}{m!P[Z_*=m]}\left(\frac{\partial^m}{\partial u^m}\right) G(\underset{\sim}{t};u)\Bigg|_{u=0} . \tag{2.4}$$

Proof. Note that

$$G(\underset{\sim}{t};u) = \prod_j \left[\sum_{k_j=0}^{M_j} p_j(k_j) u^{k_j} e^{k_j t_j}\right]$$

$$= \sum_{\ell=0}^{M_*} \left[\sum_{\underset{\sim}{k} \in S(\ell)} \prod_j p_j(k_j) e^{k_j t_j} \right] u^\ell$$

$$= \sum_{\ell=0}^{M_*} \{P[Z_* = \ell] M(\underset{\sim}{t};\ell)\} u^\ell. \tag{2.5}$$

The result follows by straightforward differentiation.

Theorem 1 can now be used to derive a useful formula for the moments of $p(\underset{\sim}{k})$.

Theorem 2. Let $\underset{\sim}{Y}$ be a random vector with distribution $p(\underset{\sim}{k})$. Then

$$E\prod_j Y_j^{\nu_j} = \frac{1}{m!P[Z_*=m]} \left(\frac{d^m}{du^m}\right) \left[\prod_j \frac{\partial^{\nu_j}}{\partial t_j^{\nu_j}} g_j\left(ue^{t_j}\right) \Bigg|_{t_j=0} \right]_{u=0} . \tag{2.6}$$

Proof. It is well known that

$$E\prod_j Y_j^{\nu_j} = \frac{\partial^{\nu_*}}{\partial t_1^{\nu_1} \cdots \partial t_r^{\nu_r}} M(\underset{\sim}{t};m) \Bigg|_{\underset{\sim}{t}=0} \tag{2.7}$$

The result is obtained by using (2.4) in (2.7) and interchanging the order of differentiation.

The advantage of (2.6) over (2.7) is that in (2.6) the partial derivatives w.r.t. the t_j's are done independently of one another. This will become more evident in Section 3, where (2.6) is applied.

3. THE COMPOUND MULTINOMIAL MOMENTS

The compound multinomial (CMtn) distribution defined in Table 1 arises in a Bayesian approach to making inferences about the parameter vector $\underset{\sim}{M}$ of the multivariate hypergeometric distribution. Hoadley [1969] shows how the moments of the CMtn distribution may be used in approximating the posterior distribution of parameters having the form $f(\underset{\sim}{M})$. The method used in Hoadley [1969] for calculating these moments is to first compute the joint factorial moments and then use conversion formulas to get the joint moments. We now use (2.6) to derive a different formula for these moments which might be better suited for efficient computation. Without loss of generality, it is assumed that $\nu_j > 0$ for $j \leq s$ and $\nu_j = 0$ for $j > s$. Of course s could equal r.

For this case $M_j = \infty$,

$$p_j(k) = \frac{\Gamma(k+a_j)}{k!\,\Gamma(a_j)} q^k (1-q)^{a_j},$$

$$P\{Z_*=m\} = \frac{\Gamma(m+a_*)}{m!\,\Gamma(a_*)} q^m (1-q)^{a_*}, \tag{3.1}$$

and

$$g_j(w) = (1-q)^{a_j}(1-qw)^{-a_j}.$$

Before applying (2.6), note that for $\ell \geq 0$

$$\frac{d^\ell}{dt^\ell}(1-que^t)^{-a} =$$

$$(1-que^t)^{-(a+\ell)}\left[\sum_{k=1}^{\ell} C_{\ell,k}(a)(qu)^k e^{kt}\right], \tag{3.2}$$

where $C_{\ell,k}(a)$ is a kth degree polynomial in a; and when $\ell = 0$, $\sum_{k=1}^{\ell}$ is taken to be 1. Use of (3.1) and (3.2) yield

$$\frac{1}{m!P\{Z_*=m\}} \prod_j \frac{d^{\nu_j}}{dt_j^{\nu_j}} g_j\left(ue^{t_j}\right)\Bigg|_{t_j=0} =$$

$$\frac{\Gamma(a_*)}{\Gamma(m+a_*)q^m} \prod_j (1-qu)^{-(a_j+\nu_j)} \sum_{k_j=1}^{\nu_j} C_{\nu_j,k_j}(a_j)(qu)^{k_j} =$$

$$\frac{\Gamma(a_*)(1-qu)^{-(a_*+\nu_*)}}{\Gamma(m+a_*)q^m} \sum_{k_1=1}^{\nu_1} \cdots \sum_{k_s=1}^{\nu_s} \prod_{j=1}^{s} C_{\nu_j,k_j}(a_j)(qu)^{k_j} =$$

$$\frac{\Gamma(a_*)(1-qu)^{-(a_*+\nu_*)}}{\Gamma(m+a_*)q^m} \sum_{w=s}^{\nu_*} A_w(qu)^w, \tag{3.3}$$

where

$$A_w = \sum_{\substack{1 \le k_j \le \nu_j \\ k_1+\dots+k_s=w}} \prod_{j=1}^{s} C_{\nu_j,k_j}(a_j).$$

By expanding $(1-qu)^{-(a_*+\nu_*)}$ in a series, (3.3) becomes

$$\frac{\Gamma(a_*)}{\Gamma(m+a_*)q^m} \sum_{w=s}^{\nu_*} A_w \sum_{k=0}^{\infty} \frac{\Gamma(k+a_*+\nu_*)}{k!\Gamma(a_*+\nu_*)}(qu)^{k+w};$$

hence,

$$E \prod_{j=1}^{s} Y_j^{\nu_j} = \frac{\Gamma(a_*)}{\Gamma(m+a_*)q^m} \sum_{w=s}^{\nu_*} A_w \left\{ \frac{d^m}{du^m} \sum_{k=0}^{\infty} \frac{\Gamma(k+a_*+\nu_*)}{k!\Gamma(a_*+\nu_*)} (qu)^{k+w} \right\} \Bigg|_{u=0}$$

$$= \begin{cases} \dfrac{m!\Gamma(a_*)}{\Gamma(m+a_*)} \displaystyle\sum_{w=s}^{\min\{\nu_*,m\}} A_w \left[\frac{\Gamma(m-w+a_*+\nu_*)}{(m-w)!\Gamma(a_*+\nu_*)} \right] & \text{if } s \le m. \\ 0 \text{ if } s > m. \end{cases} \tag{3.5}$$

That the expectation is zero when $s > m$ makes sense, because in this case, one of the Y_j's, $j = 1,\dots,s$, must be zero.

In order for this formula to be useful, the A_w's must be evaluated. To do this, one needs to know $C_{\ell,k}(a)$ for $k = 1,2,\dots,\ell$; $\ell = 1,2,\dots$. The easiest way to find the $C_{\ell,k}(a)$'s is to derive a recurrence relationship which they satisfy. If we let $v = qu$, differentiate (3.2) w.r.t. t, and set $t = 0$, the result is

$$\frac{\partial^{\ell+1}}{\partial t^{\ell+1}}(1-ve^t)^{-a} \Bigg|_{t=0}$$

$$= (1-v)^{-(a+\ell+1)} \left[(1-v) \sum_{k=1}^{\ell} C_{\ell,k}(a)kv^k + (a+\ell)v \sum_{k=1}^{\ell} C_{\ell,k}(a)v^k \right]$$

$$= (1-v)^{-(a+\ell+1)} \left\{ C_{\ell,1}(a)v + \sum_{j=2}^{\ell} \left[jC_{\ell,j}(a) + (a+\ell-j+1)C_{\ell,j-1}(a) \right] v^j + aC_{\ell,\ell}(a)v^{\ell+1} \right\}. \tag{3.6}$$

Similarly, if we replace ℓ by $\ell+1$ in (3.2) and set $t = 0$, the result is

$$\left.\frac{d^{\ell+1}}{dt^{\ell+1}}(1-ve^t)^{-a}\right|_{t=0}$$

$$= (1-v)^{-(a+\ell+1)}\left\{\sum_{k=1}^{\ell+1} C_{\ell+1,k}(a)v^k\right\}. \tag{3.7}$$

By equating coefficients of like powers in (3.6) and (3.7) we get

$$C_{\ell+1,1}(a) = C_{\ell,1}(a)$$

$$C_{\ell+1,j}(a) = jC_{\ell,j}(a) + (a+\ell-j+1)C_{\ell,j-1}(a), \quad j = 2,\ldots,\ell$$

$$C_{\ell+1,\ell+1}(a) = aC_{\ell,\ell}(a). \tag{3.8}$$

Now

$$\left.\frac{\partial}{\partial t}(1-ve^t)^{-a}\right|_{t=0} = (1-v)^{-(a+1)}(av); \tag{3.9}$$

hence $C_{1,1}(a) = a$, and the rest of the $C_{\ell,k}(a)$'s follow by applying (3.8) recursively.

It is easily shown that $C_{\ell,k}(a)$ has the form

$$C_{\ell,k}(a) = \sum_{n=1}^{k} d_{\ell,k,n}a^n.$$

It follows from (3.8) that $d_{\ell,1,1} = 1$, $d_{\ell,\ell,n} = 0$ for $n < \ell$, and $d_{\ell,\ell,\ell} = 1$. A computer program was written to evaluate the rest of the $d_{\ell,k,n}$'s for ℓ up to 10. The result appears in Table 2.

The result is stated completely in

Theorem 3. If $\underset{\sim}{Y}(1\times r) \sim CMtn(m;\underset{\sim}{a})$ and $\nu_1,\ldots,\nu_s$ are positive integers, then

$$E\prod_{j=1}^{s} Y_j^{\nu_j} = \frac{m!\Gamma(a_*)}{\Gamma(m+a_*)}\sum_{w=s}^{\min\{\nu_*,m\}} A_w\left[\frac{\Gamma(m-w+a_*+\nu_*)}{(m-w)!\Gamma(a_*+\nu_*)}\right],$$

where

$$A_w = \sum_{\substack{1 \le k_j \le \nu_j \\ k_1 + \ldots + k_s = w}} \prod_{j=1}^{s} C_{\nu_j, k_j}(a_j),$$

$$C_{\ell,k}(a) = \sum_{n=1}^{k} d_{\ell,k,n} a^n$$

and $d_{\ell,1,1} = 1$, $d_{\ell,\ell,n} = 0$ for $n < \ell$, $d_{\ell,\ell,\ell} = 1$. The rest of the $d_{\ell,k,n}$'s are given in Table 2.

TABLE 2: The $d_{\ell,k,n}$'s of Theorem 3

ℓ	k	n = 1	n = 2	n = 3	n = 4	n = 5	n = 6	n = 7	n = 8	n = 9
3	2	1	3							
4	2	4	7							
	3	1	4	6						
5	2	11	15							
	3	11	30	25						
	4	1	5	10	10					
6	2	26	31							
	3	66	146	90						
	4	26	91	120	65					
	5	1	6	15	20	15				
7	2	57	63							
	3	302	588	301						
	4	302	868	896	350					
	5	57	238	406	350	140				
	6	1	7	21	35	35	21			
8	2	120	127							
	3	1191	2136	966						
	4	2416	6126	5376	1701					
	5	1191	4096	5586	3696	1050				
	6	120	575	1176	1316	840	266			
	7	1	8	28	56	70	56	28		
9	2	247	255							
	3	4293	7290	3025						
	4	15619	36275	28470	7770					
	5	15619	47400	55560	30660	6951				
	6	4293	16929	27910	24570	11886	2646			
	7	247	1326	3123	4200	3486	1764	462		
	8	1	9	36	84	126	126	84	36	
10	2	502	511							
	3	14608	23902	9330						
	4	88234	193533	139320	34105					
	5	156190	434494	456525	220620	42525				
	6	88234	306793	437100	325620	129780	22827			
	7	14608	64362	122520	131020	84630	32172	5880		
	8	502	2971	7860	12195	12180	8022	3360	750	
	9	1	10	45	120	210	252	210	120	45

REFERENCES

[1] Hoadley, A. B. 1969. The Compound Multinomial Distribution and Bayesian Analysis of Categorical Data From Finite Populations. J. Amer. Statist. Ass. 64, 216-29.

[2] Tsao, C. K. 1965. A Moment Generating Function of the Hypergeometric Distribution. Classical and Contagious Discrete Distributions. (Ed. G. P. Patil). Statistical Publishing Society, Calcutta (Distributed outside India by Pergamon Press), 75-78.

LOCATION OF MODES FOR CERTAIN UNIVARIATE AND MULTIVARIATE DISCRETE DISTRIBUTIONS

K. G. JANARDAN
G. P. PATIL

Department of Statistics
The Pennsylvania State University
University Park, Pennsylvania

SUMMARY

A systematic survey of location of modes for various types of discrete distributions.

1. INTRODUCTION

The problem of location of modes for discrete distributions has received some attention in statistical literature in the recent past. In this paper we attempt a systematic investigation on the subject for certain univariate and multivariate discrete distributions. Several individual results have appeared before; some others are new and are on record for the first time in the present paper.

2. PRELIMINARIES

The probability function (pf) of a univariate discrete distribution is denoted by $p(x)$. The pf of a multivariate discrete distribution is denoted by $p(x_1,x_2,\ldots,x_s)$. The ratio of the pf's, $p(x+1)/p(x)$, is denoted by $R(x)$ and is called the ratio function.

To locate a mode (or modes) of a univariate discrete distribution, we use the following result: as x goes from 0 to n (n could be ∞), the terms $p(x)$ first increase monotonically, then decrease monotonically, reaching their greatest value when $x = m$, except that $p(x+1) = p(x)$ when $m_2 = m_1+1$. Equivalently, the ratio function $R(x) \leq 1$ defines the mode or modes of a univariate discrete distribution.

To locate a mode (or modes) of a multivariate discrete distribution, we adopt Moran's approach, given below, for which we shall introduce the following definitions due to [2].

a. The vector $\underset{\sim}{y} = (y_1,y_2,\ldots,y_s)$ is called [i,j] neighbor of the vector $\underset{\sim}{x} = (x_1,x_2,\ldots,x_s)$ if $y_i = x_i+1$, $y_j = x_j-1$ and $y_k = x_k$ for $k \neq i,j$.

b. $\underset{\sim}{x}$ is a strict local mode if its probability exceeds that of all of its neighbors.

c. A joint mode is one which has an equiprobable neighbor.

Moran's condition for $\underset{\sim}{x}$ to be a strict local mode: using definitions a and b, we obtain the condition as

$$\frac{p(x_1,\ldots,x_i+1,\ldots,x_j-1,\ldots,x_s)}{p(x_1,x_2,\ldots,x_s)} = N(x_i,x_j) < 1$$

for all $i,j = 1,2,\ldots,s$. To emphasize the dependence of the ratio of the probabilities on the i-th and j-th components we denote the ratio by $N(x_i, x_j)$ and call it a "Neighborhood Function."

Moran's condition for x to be a joint mode: using definitions a and c, we obtain the condition as

$$\frac{p(x_1,\ldots,x_i+1,\ldots,x_j-1,\ldots,x_s)}{p(x_1,x_2,\ldots,x_s)} = N(x_i,x_j) \leq 1$$

for all $i,j = 1,2,\ldots,s$, the equality occurring at the equally probable neighbors.

Feller gives a problem ([1], p. 161) for finding the mode of a singular multinomial distribution, which provides a fairly sharp inequality for each x_i, due to Moran. The condition $\sum_{i=1}^{s} x_i = n$ further restricts the possible modes to a few points the evaluation of the neighborhood function, as we stated above, to locate the mode or the joint modes. [2] Gives a constructive method (see Section 4) of finding the mode or the joint modes for the singular multinomial distribution and also applies it to the singular multivariate hypergeometric distribution. Here we adopt that approach to locate the mode or modes of all the singular multivariate discrete distributions satisfying the condition $\sum_{i=1}^{s} x_i \neq n$, n being finite. For the multivariate distributions where $\sum_{i=1}^{s} x_i \neq n$, we record Feller-Moran inequalities for each component x_i with the assumption that $c \leq \sum_1^s x_i \leq d$.

3. UNIVARIATE DISCRETE DISTRIBUTIONS

3.1 Binomial Distributions: $p(x) = \binom{n}{x} p^x (1-p)^{n-x}$ $\quad x = 1,2,\ldots,n$

$n = 1,2,\ldots (n < \infty)$

$0<p<1.$

$$R(x) = \frac{(n-x)}{(x+1)}\left[\frac{p}{(1-p)}\right] \gtreqless 1 \text{ according as } x \lesseqgtr (n+1)p-1$$

p(x) is bimodal when $m_1 = (n+1)p-1$ is an integer, and the modes are m_1 and $m_2 = m_1+1$. p(x) is unimodal when $m_1 = (n+1)p-1$ is not an integer and the mode is $m = [m_1+1]$, where $[m_1+1]$ is the integral part of m_1+1.

$$m_1 = \mu\left(\frac{n+1}{n}\right)-1, \text{ where } \mu = np$$

$$m_2 = m_1+1 = \mu\left(\frac{n+1}{n}\right).$$

3.2 Poisson Distribution: $p(x) = \frac{e^{-\lambda}\lambda^x}{x!}$ $\quad x = 0,1,2,\ldots$

$0<\lambda<\infty$

$$R(x) = \frac{1}{x+1}\lambda \gtreqless 1 \text{ according as } x \lesseqgtr \lambda-1$$

p(x) is bimodal when $m_1 = \lambda-1$ is an integer, and the modes are m_1 and $m_2 = m_1+1$. p(x) is unimodal when $m_1 = \lambda-1$ is not an integer and the mode is $m = [\lambda]$.

$$m_1 = \lambda-1 = \mu-1$$

$$m_2 = \lambda = \mu.$$

3.3 Displaced Poisson Distribution:

$$p(x) = \frac{e^{-\lambda}\lambda^{x+r}}{I(r,\lambda)(x+r)!} \qquad x = 0,1,2,\ldots \quad 0<\lambda<\infty \quad r = 0,1,2,\ldots$$

where $I(r,\lambda) = \sum_{y=r}^{\infty} \frac{e^{-\lambda}\lambda^y}{y!}$

$$R(x) = \frac{1}{x+r-1} \gtreqless 1 \text{ according as } x \lesseqgtr (\lambda-r-1)$$

p(x) is bimodal when $m_1 = \lambda-r-1$ is an integer and the two modes are m_1 and $m_2 = m_1+1$. p(x) is unimodal when $m_1 = \lambda-r-1$ is not an integer and the mode is $m = [m_1+1] = [\lambda-r]$.

3.4 Hyper-Poisson Distribution:

$$p(x) = \frac{\Gamma(\lambda)\theta^x}{{}_1F_1[1;\lambda;\theta]\Gamma(\lambda+x)} \qquad x = 0,1,2,\ldots \quad 0<\theta<\infty \quad 0<\lambda<\infty$$

$$R(x) = \frac{1}{\lambda+x}\theta \gtreqless 1 \text{ according as } x \lesseqgtr (\theta-\lambda).$$

Also p(x) is bimodal when $m_1 = \theta-\lambda$ is an integer, and the two modes are $m_1 = \theta-\lambda$, $m_2 = m_1+1 = \theta-\lambda+1$. p(x) is unimodal when $m_1 = \theta-\lambda$ is not an integer, and the mode is $m = [m_1+1] = [\theta-\lambda+1]$.

3.5 Negative Binomial Distribution:

$$p(x) = \binom{x+k-1}{x} p^k (1-p)^x \qquad x = 0,1,2,\ldots \quad 0<k<\infty \quad 0<p<1$$

$$R(x) = \frac{x+k}{x+1}(1-p) \gtreqless 1 \text{ according as } x \lesseqgtr \frac{(1-p)k-1}{p}$$

p(x) is bimodal when $m_1 = \frac{(1-p)k-1}{p}$ is an integer and the two modes are m_1 and $m_2 = m_1+1$. p(x) is unimodal when $m_1 = \frac{(1-p)k-1}{p}$ is not an integer and the mode is $m = [m_1+1] = \frac{(1-p)(k-1)}{p}$.

$$m_1 = \mu\left(\frac{k-1}{k}\right)-k, \text{ where } \mu = \frac{k(1-p)}{p}$$

$$m_2 = m_1+1 = \mu\left(\frac{k-1}{k}\right)$$

3.6 Pascal Distribution:

$$p(x) = \binom{x-1}{k-1}p^k q^{x-k} \qquad x = k,k+1,k+2,\ldots$$
$$k = 1,2,\ldots$$
$$0<p<1,\ q = 1-p$$

$$R(x) = \frac{x}{x-k+1}(1-p) \gtreqless 1 \text{ according as } x \lesseqgtr \frac{k-1}{p}$$

$p(x)$ is bimodal when $m_1 = \frac{k-1}{p}$ is an integer, and the two modes are m_1 and $m_2 = m_1+1 = \frac{k-(1-p)}{p}$.

$$m_1 = \mu-\frac{1}{p} \text{ where } \mu = \frac{k}{p}$$

$$m_2 = \mu-\frac{1-p}{p}$$

Form 2: $$p(x) = \binom{x+k-1}{k-1}p^k(1-p)^x \qquad x = 0,1,2\ldots$$
$$k = 1,2,\ldots$$
$$0<p<1$$

NOTE: The pascal distribution, form 2, is a special caşe of the negative binomial distribution with parameters k and p when k is an integer, and hence the mode of this distribution is the same as the mode of the negative binomial distribution when k is an integer.

3.7 Logarithmic Distribution:

$$p(x) = \frac{\alpha\theta^x}{x} \qquad x = 1,2,\ldots$$
$$0<\theta<1,$$

where $\alpha = -[\log(i-\theta)]^{-1}$.

$$R(x) = \frac{x}{x+1}\theta \gtreqless 1 \text{ according as } x \lesseqgtr \frac{1}{\theta-1}.$$

$p(x)$ is unimodal with mode m = g.l.b. $\{1,2,\ldots\} = 1$.

3.8 Polya-Eggenberger Distribution:

$$p(x) = \frac{\Gamma(x+\frac{h}{\theta})}{x!\Gamma(\frac{h}{\theta})}\left(\frac{1}{1+\theta}\right)^{h/\theta}\left(\frac{1}{1+\theta}\right)^x \qquad x = 0,1,2,\ldots$$
$$0<h<\infty$$
$$0<\theta<\infty$$

$$R(x) = \left(\frac{x+\frac{h}{\theta}}{x+1}\right)\left(\frac{\theta}{1+\theta}\right) \gtreqless 1 \text{ according as } x \lesseqgtr h-1-\theta.$$

$p(x)$ is bimodal when $m_1 = h-1-\theta$ is an integer and the two modes are m_1 and $m_2 = m_1+1 = h-\theta$. And $p(x)$ is unimodal when m_1 is not an integer and the mode is $m = [m_1+1] = [h-\theta]$.

NOTE: If $h-\theta$ is negative, mode $m = 0$.

3.9 Modified Logarithmic Distribution:

$$p(x) = \begin{cases} \delta & \text{when } x = 0 \quad 0 \le \delta < 1 \\ 1-\delta \dfrac{\alpha\theta^x}{x} & \text{when } x = 1,2,\ldots \quad 0<\theta<1 \end{cases}$$

where $\alpha = -[\log(1-\theta)]^{-1}$.

$$R(x) = \frac{x}{x+1} \gtreqless 1 \text{ according as } x \lesseqgtr \frac{1}{\theta-1}$$

$p(x)$ is unimodal with mode m = g.l.b. $\{1,2,\ldots\} = 1$.

3.10 Power Series Distribution:

$$p(x) = \frac{a(x)\theta^x}{f(\theta)}, \quad x \in T, \tag{3.1}$$

where T is a subset of the non-negative integers, the series function $f(\theta)$ is given by $f(\theta) = \Sigma a(x)\theta^x$, the summation extending over all $x \in T$ such that $a(x) > 0$ and $0 \le \theta < R$, where R is the radius of convergence of the series function $f(\theta)$. Mode of PSD is defined by

$$R(x) \lesseqgtr 1 \iff \theta a(x+1) \gtreqless a(x)$$

Theorem 3.10.1:

If the ratio $R' = \frac{a(x+1)}{a(x)}$ of the coefficients of the series function is the quotient of two linear functions of x of the form $\frac{ax+b}{x+c}$, where a,b,c are real numbers such that $\frac{b\theta-c}{1-a\theta} > 0$, then equation 3.1 is bimodal with modes $m_1 = \frac{b\theta-c}{1-a\theta}$ and $m_2 = m_1+1$ if m_1 is integral, and it is unimodal with mode $m = [m_1+1]$ when m_1 is not an integer.

NOTE: Each of m_1 and m_2 is meaningful only if it is ε T.

Proof: $R(x) = \frac{a(x+1)}{a(x)}\theta$; let $\frac{a(x+1)}{a(x)} = \frac{ax+b}{x+c}$, where a, b, c are real numbers such that $\frac{b\theta-c}{1-a\theta} > 0$. $R(x) = \frac{ax+b}{x+c}\theta \lesseqgtr 1$ according as $x \gtreqless \frac{b\theta-c}{1-a\theta}$. Hence $p(x)$ is bimodal when $m_1 = \frac{b\theta-c}{1-a\theta}$ is an integer and the two modes are m_1 and $m_2 = m_1+1$. $P(x)$ is unimodal when $m_1 = \frac{b\theta-c}{1-a}$ is not an integer and the mode $m = [m_1+1]$.

Theorem 3.10.2:

If the ratio $\frac{a(x+1)}{a(x)}$ of the coefficients of the series function is the quotient of two linear functions of x of the form $\frac{ax+b}{x+c}$ where a,b,c are real numbers such that $\frac{b\theta-c}{1-a\theta} \leq 0$, then the PSD eq. 3.1, is bimodal with modes $m_1 = 0$ and $m_2 = 1$ when $\frac{b\theta-c}{1-a\theta} = 0$ and is unimodal with mode m = g.l.b.(T) if $\frac{b\theta-c}{1-a\theta} < 0$.

Proof: Let $\frac{a(x+1)}{a(x)} = \frac{ax+b}{x+c}$, where a,b,c are real, such that $\frac{b\theta-c}{1-a\theta} \leq 0$. $R(x) = \frac{ax+b}{x+c}\theta \gtreqless 1$ according as $x \lesseqgtr \frac{b\theta-c}{1-a\theta}$. If $m_1 = \frac{b\theta-c}{1-a\theta} = 0$, then $x = 0$ and $p(x+1) = p(x)$; hence the PSD is bimodal with modes $m_1 = 0$ and $m_2 = 1$. If $\frac{b\theta-c}{1-a\theta} < 0$, the $p(x+1) < p(x)$, and so the PSD is unimodal with m = g.l.b.(T).

REMARKS: Binomial, Poisson, and negative binomial distributions are PSD's with the coefficients of the series function and parametric relations as given in the table below. The modes of these distributions can be found using Theorem 3.10.1 and the table. Similarly, logarithmic distribution is a PSD which satisfies Theorem 3.10.2, and so the mode can be found accordingly.

Distribution	Parameter	Reparametrisation	Coefficient	Ratio	a	b	c
1 Binomial	$0<p<1$	$\theta = \frac{p}{1-p}$	$\binom{n}{x}$	$\frac{n-x}{x+1}$	-1	n	1
2 Poisson	$0<\lambda<\infty$	$\theta = \lambda$	$\frac{1}{x!}$	$\frac{1}{x+1}$	0	1	1
3 Negative binomial	$0<k<\infty$ $0<p<1$	$\theta = 1-p$	$\binom{x+k+1}{x}$	$\frac{x+k}{x+1}$	1	k	1
4 Logarithmic	$0<\theta<1$	$\theta = \theta$	$\frac{1}{x}$	$\frac{x}{x+1}$	1	0	·1

3.11 Hypergeometric Distribution:

$$p(x) = \frac{\binom{M_1}{x}\binom{N-M_1}{n-x}}{\binom{N}{n}}$$

N, M_1, and n are positive integers, $N>M_1>n$, $x = \max(0, n-N+N_1)\ldots, \min(n,M_1)$.

$$R(x) = \frac{M_1(n-x)-nx+x^2}{N(x+1)-M_1(x+1)-n(x+1)+x+1)^2} \gtreqless 1$$

according as $x \lesseqgtr \frac{(n+1)M_1-N+n-1}{N+2}$

p(x) is bimodal when $m_1 = \frac{(n+1)M_1-N+n-1}{N+2}$ is an integer and the two modes are m_1 and $m_2 = m_1+1$. p(x) is unimodal when $m_1 = \frac{(n+1)M_1-N+n-1}{N+2}$ is not an integer and the unique mode is $m = [m_1+1]$.

3.12 Inverse Hypergeometric Distribution:

$$p(x) = \frac{\binom{x-1}{k-1}\binom{N-x}{M-k}}{\binom{N}{M}} \qquad \begin{aligned} x &= k,k+1,k+2,\dots,N-M+k \\ N &= 1,2,\dots \\ M &= 1,2,\dots;\ M < N \\ k &= 1,2,\dots,M \end{aligned}$$

$$R(x) = \frac{k}{x-k+1}\left(\frac{N-x-M+k}{N-x}\right) \gtreqless 1 \text{ according as } x \lesseqgtr \frac{N(k-1)}{(M-1)}.$$

p(x) is bimodal when $m_1 = \frac{N(k-1)}{M-1}$ is an integer and the two modes are m_1 and $m_2 = m_1+1$. p(x) is unimodal when $m_1 = \frac{N(k-1)}{M-1}$ is not an integer and the mode is $m = [m_1+1] = \frac{Nk-N+M-1}{M-1}$.

3.13 Negative Hypergeometric Distribution:

$$p(x) = \frac{\binom{-M}{x}\binom{-N+M}{n-x}}{\binom{-N}{n}} \qquad \begin{aligned} x &= 0,1,2,\dots,n \\ &0<N<\infty \\ &0<M<N \\ n &= 1,2,\dots \end{aligned}$$

$$R(x) = \frac{-Mn+(M-n)x+x^2}{-N(x+1)+M(x+1)-n(x+1)+(x+1)^2} \gtreqless 1,$$

$$\text{according as } x \lesseqgtr \frac{-M(n+1)+n+N-1}{(-N+2)}$$

p(x) is bimodal when $m_1 = \frac{-M(n+1)+n+N-1}{(-N+2)}$ is an integer and the two modes are m_1 and $m_2 = m_1+1$. p(x) is unimodal when m_1 is not an integer with the mode $m = [m_1+1]$.

3.14 Polya Distribution: $p(x) = \binom{n}{x}\frac{a^{(x,c)}b^{(n-x,c)}}{(a+b)^{(n,c)}}$ where, for example, $a^{(b,c)} = a(a+c)\dots[a+(b-1)c]$, $x = 0,1,2,\dots,n$.

$$R(x) = \left(\frac{n-x}{x+c}\right)\frac{a+xc}{b+(n-x-1)c} \gtreqless 1, \text{ according as } x = \frac{n(a-c)-b+c}{a+b-2c}.$$

The Polya distribution is bimodal when $m_1 = \frac{n(a-c)-b+c}{a+b-2c}$ is an integer, with the two modes m_1 and $m_2 = m_1+1$. The Polya distribution is unimodal when $m_1 = \frac{n(a-c)-b+c}{a+b-2c}$ is not an integer and the unique mode is $m = [m_1+1]$.

NOTE: When $c = -1$, the mode of the Polya distribution reduces to the mode of the hypergeometric distribution.

3.15 Inverse Polya Distribution:

$$p(x) = \binom{k+x-1}{x} a^{(k,c)} b^{(x,c)} / (a+b)^{(x+k,c)}$$

$x = 0,1,2,\ldots \qquad c = 1,2,\ldots$

$a = 1,2,\ldots \qquad k = 1,2,\ldots$

$b = 1,2,\ldots$

$$R(x) = \left(\frac{k+x}{x+1}\right) \frac{b+xc}{(a+b)+(k+x)c} \gtreqless 1 \text{ according as } x \lesseqgtr \frac{k(b-c)-(a+b)}{(a+c)}.$$

$p(x)$ is bimodal with modes m_1 and $m_2 = m_1+1$ when $m_1 = \frac{k(b-c)-(a+b)}{(a+c)}$ is an integer. $p(x)$ is unimodal with unique mode $m = [m_1+1]$ when $m_1 = \frac{k(b-c)-(a+b)}{(a+c)}$ is not an integer.

3.16 Family of Generalized Hypergeometric Distributions:

$$p(x) = \frac{\binom{a}{x}\binom{b}{n-x}}{\binom{a+b}{n}}, \qquad x = 0,1,2,\ldots,$$

where a, b, and c are as defined in [5].

$$R(x) = \frac{na-(a+n)x+x^2}{b(x+1)-n(x+1)+(x+1)^2} \gtreqless 1 \text{ according as } x \lesseqgtr \frac{na+n-b-1}{a+b+2}.$$

$p(x)$ is bimodal with modes m_1 and $m_2 = m_1+1$ when $m_1 = \frac{na+n-b-1}{a+b+2}$ is an integer. $p(x)$ is unimodal with unique mode $m = [m_1+1]$ when $m_1 = \frac{na+n-b-1}{a+b+2}$ is not an integer.

3.17 Distribution of Exceedances:

$$p(x) = \frac{\binom{-M}{x}\binom{-N-1+m}{n-x}}{\binom{-N-1}{n}}$$

$x = 0,1,\ldots,n$

$n = 0,2,\ldots;\ n < \infty$

$N = 1,2,\ldots;\ N < \infty$

$M = 1,2,\ldots,N$

$$R(x) = \left(\frac{-M-x}{x+1}\right) \frac{(n-x)}{(-N-1+M-n+x+1)} \gtreqless 1 \text{ according as}$$

$$x \lesseqgtr \frac{M(n+1)-N-n}{N-1}.$$

$p(x)$ is bimodal when $m_1 = \frac{M(n+1)-N-n}{N-1}$ is an integer and the two modes are m_1 and $m_2 = m_1+1$. $p(x)$ is unimodal when $m_1 = \frac{M(N+1)-N-n}{N-1}$

is not an integer and the mode is $m = [m_1+1]$.

3.18 Ising Stevens Distribution:

$$p(x) = \frac{\binom{n_1-1}{x-1}\binom{n_2}{x}}{\binom{n_1+n_2-1}{n_1}}, \quad x = 1,2,\ldots.$$

$$R(x) = \frac{(n_1-x)(n_2-x)}{x(x+1)} \gtreqless 1 \text{ according as } x = \frac{n_1n_2}{n_1+n_2+1}.$$

$p(x)$ is bimodal when $m_1 = \frac{n_1n_2}{n_1+n_2+1}$ is an integer and the two modes are m_1 and $m_2 = m_1+1$. $p(x)$ is unimodal when $m_1 = \frac{n_1n_2}{n_1+n_2+1}$ is not an integer and the mode is $m = [m_1+1]$.

3.19 Leo Katz Class of Discrete Probability Distributions:

$$\frac{f(x+1)}{f(x)} = \frac{\alpha+\beta x}{x+1}, \qquad x = 0,1,2,\ldots \quad \alpha > 0; \quad \beta < 1$$

where $f(x) = p(X=x) \geq 0$, $\sum_0^\infty f(x) = 1$.

$$R(x) = \frac{\alpha+\beta x}{x+1} \gtreqless 1 \text{ according as } x \lesseqgtr \frac{1-\alpha}{\beta-1}.$$

$f(x)$ is bimodal with modes m_1 and $m_2 = m_1+1$ when $m_1 = \frac{1-\alpha}{\beta-1}$ is an integer. $f(x)$ is unimodal with mode $m = [m_1+1]$ when $m_1 = \frac{1-\alpha}{\beta-1}$ is not an integer.

4. SINGULAR MULTIVARIATE DISCRETE DISTRIBUTIONS

4.1 Singular Multinomial Distribution: $p(x_1,x_2,\ldots,x_s) = \frac{n!}{x_1!x_2!\ldots x_s!} p_1^{x_1}p_2^{x_2}\ldots p_x^{x_s}$, where $x_i = 0,1,2,\ldots;n$ for $i = 1,2,\ldots,s$, so that $\sum_{i=1}^{s} x_i = n$; $0 \leq p_i \leq 1$ for $i = 1,2,\ldots,s$, so that $\sum_{i=1}^{s} p_i = 1$.

$$N(x_i,x_j) < 1 \Rightarrow \frac{x_i+1}{p_i} > \frac{x_j}{p_j} \text{ for all } i,\ j = 1,2,\ldots,s \tag{4.1}$$

$$N(x_i,x_j) \leq 1 \Rightarrow \frac{x_i+1}{p_i} > \frac{x_j}{p_j} \text{ for all } i,\ j = 1,2,\ldots,s \tag{4.2}$$

equality occurring at equally probable neighbors.

Finucan's condition for $\underset{\sim}{x}$ to be a strict local mode.

If there exists a set

$$A = \{N \mid \sum_{i=1}^{s} [Np_i] = n\},$$ where $[Np_i]$ denotes the integral part of Np_i, then each x_i is uniquely defined by $x_i = [Np_i]$ such that $\sum_{i=1}^{s} x_i = n$ for any $N \varepsilon A$ and $\underset{\sim}{x}$ is the unique mode.

Finucan's conditions for $\underset{\sim}{x}$ to be a joint mode:

If there exists a single N for which $r \geq 2$ of the Np_i's are integers, then there exists a set

$$B = \left\{\underset{\sim}{m} \mid \sum_{i=1}^{s} m_i = n, \text{ where } m_i = [Np_i]\right\}$$ and any $\underset{\sim}{x} \varepsilon B$ is a joint mode.

That the conditions in equations 4.1 and 4.2 of Moran are equivalent to Finucan's conditions, above, is shown in [2], from which we have the following result.

The mode(s) may be obtained by finding a number N such that the s-quantities $x_i = [NP_i]$ satisfy $\sum_{i=1}^{s} x_i = n$. If N is a unique number, then there are several points $\underset{\sim}{x}$ satisfying the conditions $x_i \mp [Np_i]$ and $\sum_{i=1}^{s} x_i = n$, which form a set of joint modes. If there are several N's such that $x_i = [np_i]$ and $\sum_{i=1}^{s} x_i = n$, then $\underset{\sim}{x}$ is a unique mode.

4.2 Singular Multivariate Hypergeometric Distribution:

$$p(x_1,x_2,\ldots,x_s) = \frac{\binom{n_1}{x_1}\binom{n_2}{x_2}\cdots\binom{n_s}{x_s}}{\binom{N}{n}},$$

where $x_i = 0,1,2,\ldots,N_i$ for $i = 1,2,\ldots,s$, so that $\sum_{i=1}^{s} x_i = n$; $N_i = 1,2,\ldots,N$; and $\sum_{i=1}^{s} N_i = N$.

$$\text{(a)}\quad N(x_i,x_j) < 1 \Rightarrow \frac{x_i+1}{N_i+1} > \frac{x_j}{N_j+1} \text{ for all } i,j = 1,2,\ldots,s.$$

$$\text{(b)}\quad N(x_i,x_j) \leq 1 \Rightarrow \frac{x_i+1}{N_i+1} > \frac{x_j}{N_j+1} \text{ for all } i,j = 1,2,\ldots,s.$$

The mode(s) may be obtained by finding a constant C such that

The mode(s) may be obtained by finding a constant K so that x is defined by $x_i = [K(N_i-c]$ such that $\sum_{i=1}^{s} x_i = n$. If K is unique and if $\sum_{i=1}^{s} x_i = n$, then there are many acceptable points $\underset{\sim}{x}$ which form a set of joint modes. If K is not unique, then we get a unique $\underset{\sim}{x}$ such that $x_i = [K(N_i-c)]$ and $\sum_{i=1}^{s} x_i = n$, which would be the unique mode.

4.5 Singular Sum-Symmetric Multivariate Power Series Distribution:

$$p(x_1,x_2,\ldots,x_s) = \binom{n}{\underset{\sim}{x}} \frac{a(x_1+x_2+\ldots+x_s)\prod_{i=1}^{s}\theta_i^{x_i}}{g(\theta_1+\theta_2+\ldots+\theta_s)},$$

where $(x_1,x_2,\ldots x_s) \in T$, a subset of the set of s-tuples consisting of non-negative integers and $g(\theta_1+\theta_2+\ldots+\theta_s) = \sum_S b(x_1+x_2+\ldots+x_s)$ $(\theta_1+\theta_2+\ldots+\theta_s)^n$ where $S = \left\{x_1+x_2+\ldots+x_s = n \mid (x_1,x_2\ldots,x_s \in T)\right\}$ and $b(x_1+x_2+\ldots+x_s) > 0$ for $x_1+x_2+\ldots+x_s \in S$. (See [4].)

$$N(x_i,x_j) < 1 \Rightarrow \frac{x_i+1}{\theta_i} > \frac{x_j}{\theta_j} \text{ for all } i,j = 1,2,\ldots,s.$$

$$N(x_i,x_j) \le 1 \Rightarrow \frac{x_i+1}{\theta_i} \ge \frac{x_j}{\theta_j} \text{ for all } i,j = 1,2,\ldots,s.$$

The mode(s) may be obtained by finding a constant C so that $\underset{\sim}{x}$ is defined by $x_i = [C\theta_i]$ such that $\sum_{i=1}^{s} x_i = n$. If C is unique, then there are many acceptable points $\underset{\sim}{x}$ such that $x_i = [C\theta_i]$ and $\sum_{i=1}^{s} x_i = n$ which form a set of joint modes. If there are several C's such that $x_i = [C\theta_i]$ and $\sum_{i=1}^{s} x_i = n$, then $\underset{\sim}{x}$ is unique mode of the distribution.

4.6 Unified Multivariate Hypergeometric Distribution:

$$p(x_1,x_2,\ldots,x_s) = \frac{\binom{a_1}{x_1}\binom{a_2}{x_2}\cdots\binom{a_s}{x_s}}{\binom{a}{n}},$$

where $x_1+x_2+\ldots+x_s = n$ and $a_1+a_2+\ldots+a_s = a$ (from [3]).

$$N(x_i,x_j) < 1 \Rightarrow \frac{x_i+1}{a_i+1} > \frac{x_j}{a_j+1} \text{ for all } i,j = 1,2,\ldots,s.$$

$$N(x_i,x_j) \le 1 \Rightarrow \frac{x_i+1}{a_i+1} \ge \frac{x_j}{a_j+1} \text{ for all } i,j = 1,2,\ldots,s.$$

the s-quantities $x_i = [C(N_i+1]$ satisfy $\sum_{i=1}^{s} x_i = n$. If C is unique, then there are several points $\underset{\sim}{x}$, such that $x_i = [C(N_i+1]$ and $\sum_{i=1}^{s} x_i = n$, which form a set of joint modes. If there are several C's such that $x_i = [C(N_i+1)]$ and $\sum_{i=1}^{s} x_i = n$, then $\underset{\sim}{x}$ is a unique mode.

4.3 Singular Multivariate Negative Hypergeometric Distribution:

$$p(x_1,x_2,\ldots,x_s) = \frac{\binom{-N_1}{x_1}\binom{-N_2}{x_1}\cdots\binom{-N_s}{x_s}}{\binom{-N}{n}}$$

where $x_i = 0,1,2,\ldots,n$, so that $\sum_{i=1}^{s} x_i = n$, $\sum_{i=1}^{s} N_i = N$.

(a) $N(x_i,x_j) < 1 \Rightarrow \frac{x_i+1}{N_i-1} > \frac{x_i}{N_j-1}$ for all $i,j = 1,2,\ldots,s$.

(b) $N(x_i,x_j) \leq 1 \Rightarrow \frac{x_i+1}{N_i-1} \geq \frac{x_i}{N_i-1}$ for all $i,j = 1,2,\ldots,s$.

The mode(s) may be found by determining a constant C such that the s-quantities $x_i = [C(N_i-1)]$ satisfy $\Sigma x_i = n$. If C is unique then there are several $\underset{\sim}{x}$, satisfying the conditions $x_i = [C(N_i-1)]$ and $\sum_{i=1}^{s} x_i = n$, which form a set of joint modes. If there are several C's such that $x_i = [C(N_i-1)]$ and $\Sigma x_i = n$, then x is a unique mode.

4.4 Singular Multivariate Polya Distribution:

$$p(x_1,x_2,\ldots,x_s) = \binom{n}{\underset{\sim}{x}} \frac{\prod_{i=1}^{s} N_i^{(x_i,c)}}{N^{(n,c)}},$$

where $a^{(b,c)} = a(a+c)\ldots[a+(b-1)c]$, and $x_i = 0,1,2,\ldots,n$ for $i = 1,2,\ldots,s$ so that $\sum_{i=1}^{s} x_i = n$; $\sum_{i=1}^{s} N_i = N$.

$$\binom{n}{\underset{\sim}{x}} = \frac{n!}{x_1!x_2!\ldots x_s!}$$

$$N(x_i,x_j) < 1 \Rightarrow \frac{x_i+1}{\frac{N_i}{c}-1} > \frac{x_j}{\frac{N_j}{c}-1} \text{ for all } i,j = 1,2,\ldots,s,$$

where c = 0.

$$N(x_i,x_j) \leq 1 \Rightarrow \frac{x_i+1}{\frac{N_i}{c}-1} \geq \frac{x_j}{\frac{N_j}{c}-1} \text{ for all } i,j = 1,2,\ldots,s.$$

The mode(s) may be obtained by finding a constant K so that $\tilde{x}$ is defined by $x_i = [K(a_i+1)]$ and $\sum_{i=1}^{s} x_i = n$. If K is unique, then there are many acceptable points $\underset{\sim}{x}$ such that $x_i = [K(a_i+1)]$ and $\sum_{i=1}^{s} x_i = n$, which form a set of joint modes. If K is not unique, then we get a unique $\underset{\sim}{x}$ such that $x_i = [k(a_i+1)]$ and $x_i = n$ which would be the unique mode of the distribution.

5.   NON-SINGULAR MULTIVARIATE DISCRETE DISTRIBUTIONS

In this section, we record Feller-Moran inequalities for each component x_i of the mode $\underset{\sim}{x}$ of certain well-known multivariate discrete distributions which are truncated such that $c \leq \sum_{i=1} x_i \leq d$, where c and d are positive constants.

5.1 Multinomial Distribution:

$$p(x_1,x_2,\ldots,x_s) = \frac{n!}{\prod\limits_{i=0}^{s} x_i!} \prod_{i=0}^{s} p_i^{x_i} \tag{5.1}$$

where $x_i = 0,1,2,\ldots,s$ for $i = 0,1,2,\ldots,s$ and $\sum_{i=1}^{s} x_1 \leq 1$; $0 \leq p_i \leq 1$ for $i = 0,1,2,\ldots,s$ and $\sum_{i=1}^{s} p_i \leq 1$.

The mode $\underset{\sim}{x} = (x_1,x_2,\ldots,x_s)$ of the s-variate multinomial distribution (5.1) satisfies the Feller-Moran inequalities:

$$cp_i-1 < x_i \leq (d+s)p_i \text{ for } i = 1,2,\ldots,s,$$

where $c \leq \sum_{i=1}^{s} x_i \leq d$.

5.2 Negative Multinomial Distribution:

$$p(x_1,x_2,\ldots x_s) = \frac{(k+x-1)!}{(k-1)! \prod\limits_{i=1}^{s} x_i!} p_o^k \prod_{i=1}^{s} p_i^{x_i}, \tag{5.2}$$

where $x_i = 0,1,2,\ldots,\infty$; $x = \sum_{i=1}^{s} x_i$; $0<k<\infty$; $0 \leq p_i \leq 1$ for $i = 0,1,2,\ldots,s$ so that $\sum_{i=1}^{s} p_i \leq 1$.

The mode $\underset{\sim}{x} = (x_1,x_2,\ldots x_s)$ of the s-variate negative multinomial distribution (5.2) satisfies the Feller-Moran inequalities:

$$c\delta_i-1 < x_i \leq (d+s)\delta_i \text{ for } i = 1,2,\ldots,s \text{ where } \delta_i = \frac{p_i}{\prod\limits_{i=1}^{s} p_i}$$

and $c \leq \sum_{i=1}^{s} x_i \leq d$.

5.3 Multivariate Hypergeometric Distribution:

$$p(x_1,x_2,\ldots,x_s) = \frac{\prod_{i=0}^{s}\binom{N_i}{x_i}}{\binom{N}{n}}, \tag{5.3}$$

where $x_i = 0,1,2,\ldots,N_i$ for $i = 0,1,2,\ldots,s$ and $\sum_{i=1}^{s} x_i = n$; $\sum_{i=1}^{s} N_i = N$.

The mode $\underset{\sim}{x} = (x_1,x_2,\ldots,x_s)$ of the s-variate hypergeometric distribution (5.3) satisfies the Feller-Moran inequalities:

$$c\delta_i - 1 < x_i \leq (d+s)\delta_i \text{ where } \delta_i = \frac{N_i+1}{\sum_{i=1}^{s} N_i + s}, \text{ and } c \leq \sum_{i=1}^{s} x_i \leq d.$$

5.4 Multivariate Negative Hypergeometric Distribution:

$$p(x_1,x_2,\ldots,x_s) = \frac{\prod_{i=0}^{s}\binom{-N_i}{x_i}}{\binom{-N}{n}}, \tag{5.4}$$

where $x_i = 0,1,2,\ldots n$ for $i = 0,1,2,\ldots,s$; $\sum_{i=1}^{s} x_i \leq n$; and $\sum_{i=0}^{s} N_i = N$.

The mode $\underset{\sim}{x} = (x_1,x_2,\ldots x_s)$ of the s-variate negative hypergeometric distribution (5.4) satisfies the Feller-Moran inequalities:

$$c\delta_i - 1 < x_i \leq (d+s)\delta_i,\ i = 1,2,\ldots,s, \text{ where } \delta_i = \frac{N_i+1}{\sum_{i=1}^{s} N_i + s}$$

and $c \leq \sum_{i=1}^{s} x_i \leq d$.

5.5 Multivariate Inverse Hypergeometric Distribution:

$$p(x_1,x_2,\ldots,x_s) = \frac{\binom{-N-1}{-k-x}\prod_{i=1}^{s}\binom{N_i}{x_i}}{\binom{-N_o-1}{-k}}, \tag{5.5}$$

where $x_i = 0,1,2,\ldots,N_i$ for $i = 1,2,\ldots,s$; $\sum_{i=1}^{s} x_i = x$;

$N_o = N-N_1-\ldots-N_s$; $k = 1,2,\ldots,N_o$.

The mode $\underset{\sim}{x} = (x_1,x_2,\ldots,x_s)$ of the s-variate inverse hypergeometric distribution (5.5) satisfies the Feller-Moran inequalities:

$$c\delta_i - 1 < x_i \leq (d+s)\delta_i,\ i = 1,2,\ldots,s, \text{ where}$$

$$\delta_i = \frac{N_i+1}{\sum_{i=1}^{s} N_i+s} \text{ and } c \leq \sum_{i=1}^{s} x_i \leq d.$$

5.6 Multivariate Negative Inverse Hypergeometric Distribution:

$$p(x_1,x_2,\ldots,x_s) = \frac{\binom{N-1}{-k-x} \prod_{i=1}^{s} \binom{-N_i}{x_i}}{\binom{N_o-1}{-k}}, \tag{5.6}$$

where $x_i = 0,1,2,\ldots,\infty$ for $i = 1,2,\ldots,s$; $\sum_{i=1}^{s} x_i = x$.

$N = \sum_{i=0}^{s} N_i$; $0<k<\infty$. (See [3]).

The mode $x = (x_1,x_2,\ldots,x_s)$ of the s-variate negative inverse hypergeometric distribution (5.6) satisfies the Feller-Moran inequalities:

$$c\delta_i-1 < x_i \leq (d+s)\delta_i, \ i = 1,2,\ldots,s, \text{ where}$$

$$\delta_i = \frac{N_i-1}{\sum_{i=1}^{s} N_i-s} \text{ and } c\leq\Sigma x_i\leq d.$$

5.7 Multivariate Polya Distribution:

$$p(x_1,x_2,\ldots,x_s) = \frac{n!}{\prod_{i=0}^{s} x_i!} \cdot \frac{N_o^{(x_o,c)} \prod_{i=1}^{s} N_i^{(x_i,c)}}{N^{(n,c)}} \tag{5.7}$$

where $x_i = 0,1,2,\ldots,n$ for $i = 0,1,2,\ldots,s$ and $\sum_{i=1}^{s} x_i \leq n$;

$\sum_{i=1}^{s} N_i \leq N$.

The mode $\underset{\sim}{x} = (x_1,x_2,\ldots,x_s)$ of the s-variate Polya distribution (5.7) satisfies the Feller-Moran inequalities:

$$b\delta_i-1 < x_i \leq (d+s)\ \delta_i, \ i = 1,2,\ldots,s, \text{ where}$$

$$\delta_i = \frac{N_i-c}{\sum_{i=1}^{s} N_i-cs}, \text{ and } b \leq \sum_{i=0}^{s} x_i \leq d.$$

5.8 Multivariate Inverse Polya Distribution:

$$p(x_1,x_2,\ldots,s) = \frac{(k+x-1)!}{(k-1)!\prod_{i=1}^{s} x_i!} \cdot \frac{N_o^{(k,c)}\prod_{i=1}^{s} N_i^{(x_i,c)}}{N^{(k+x,c)}}, \tag{5.8}$$

where $c \neq 0$, $x_i = 0,1,2,\ldots$ for $i = 1,2,\ldots,s$; $x = \sum_{i=1}^{s} x_i$, $\sum_{i=1}^{s} N_i = N$, $0<k<\infty$ (See [3]).

The mode $\underset{\sim}{x} = (x_1,x_2,\ldots,x_s)$ of the s-variate Inverse Polya distribution (5.8) satisfies the Feller-Moran inequalities:

$$b\delta_i-1 < x_i \leq (d+s)\delta_i, \quad i = 1,2,\ldots,s, \text{ where}$$

$$\delta_i = \frac{N_i-c}{\sum_{i=1}^{s} N_i-cs}, \text{ and } b\leq\Sigma x_i\leq d.$$

5.9 Multiple Poisson Distribution:

$$p(x_1,x_2,\ldots,x_s) = e^{-(\lambda_1+\lambda_2+\ldots+\lambda_s)}\prod_{i=1}^{s}\left(\frac{\lambda_i^{x_i}}{x_i!}\right), \tag{5.9}$$

where $x_i = 0,1,2,\ldots$ for $i = 1,2,\ldots,s$; $0<\lambda_i<\infty$ for $i = 1, 2,\ldots,s$.

The mode $\underset{\sim}{x} = (x_1,x_2,\ldots,x_s)$ of the s-variate multiple Poisson distribution (5.9) satisfies the Feller-Moran inequalities:

$$c\delta_i-1 < x_i \leq (d+s)\delta_i, \quad i = 1,2,\ldots,s, \text{ where}$$

$$\delta_i = \frac{\lambda_i}{\sum_{i=1}^{s}\lambda_i} \text{ and } c\leq\Sigma x_i\leq d.$$

5.10 Multivariate Logarithmic Distribution:

$$p(x_1,x_2,\ldots,x_s) = \frac{(x_1+x_2+\ldots+x_s-1)!}{x_1!x_2!\ldots x_s!} \cdot \frac{\prod_{i=1}^{s} p_i^{x_i}}{[-\log(1-p_i-\ldots-p_s)]}, \tag{5.10}$$

where $x_i = 0,1,2,\ldots,\infty$, $0 < x_1+x_2+\ldots x_s \leq \infty$, $0 <p_i<1$ for $i = 1,2,\ldots,s$.

The mode $\underset{\sim}{x} = (x_1,x_2,\ldots,x_s)$ of the s-variate logarithmic distribution (5.10) satisfies the Feller-Moran inequalities:

$$c\delta_i - 1 < x_i \leq (d+s)\delta_i, \quad i = 1,2,\ldots,s, \text{ where}$$

$$\delta_i = \frac{p_i}{\sum_{i=1}^{s} p_i} \text{ and } c \leq \sum_{i=1}^{s} x_i \leq d.$$

5.11 Sum-Symmetric Multivariate Power Series Distribution (SSPSD):

$$P(x_1,x_2,\ldots,x_s) = \frac{(x_1+x_2+\ldots+x_s)!}{x_1!x_2!\ldots x_s!} \cdot \frac{a(x_1+x_2+\ldots+x_s)}{g(\theta_1+\theta_2+\ldots+\theta_s)} \text{ mult. by}$$

$$\theta_1^{x_1}\theta_2^{x_2}\ldots\theta_s^{x_s}, \tag{5.11}$$

where $(x_1,x_2,\ldots,x_s) \in T$, T being a subset of the set of all s-tuples consisting of non-negative integers, and $g(\theta_1+\theta_2+\ldots+\theta_s) =$

$$\sum_{S} b(x_1+\ldots+x_s)(\theta_1+\theta_2+\ldots+\theta_s)^{(x_1+x_2+\ldots+x_s)}, \text{ where}$$

$S = [x_1+x_2+\ldots+x_s \mid x_1,x_2,\ldots,x_s) \in T]$ and $b(x_1+x_2+\ldots+x_s) > 0$. See [4].

The mode $\underset{\sim}{x} = (x_1,x_2,\ldots,x_s)$ of the s-variate SSPSD (5.11) with the restriction $c \leq \sum_{i=1}^{s} x_i \leq d$ satisfies the Feller-Moran inequalities:

$$c\delta_i - 1 < x_i \leq (d+s)\delta_i, \quad i = 1,2,\ldots,s, \text{ where}$$

$$\delta_i = \frac{\theta_i}{\sum_{i=1}^{s} \theta_i} \text{ and } c \leq \sum_{i=1}^{s} x_i .$$

5.12 Unified Multivariate Hypergeometric Distribution:

$$p(x_1,x_2,\ldots,x_s) = \frac{\binom{a_o}{n-x}\prod_{i=1}^{s}\binom{a_i}{x_i}}{\binom{a}{n}}, \tag{5.12}$$

where $x_i = 0,1,2,\ldots$; $x = \sum_{i=1}^{s} x_i$, $\sum_{i=1}^{s} a_i \leq a$, $\sum^{s} x_i \leq n$ (from [3]).

The mode $\underset{\sim}{x} = (x_1,x_2,\ldots,x_s)$ of the s-variate unified multivariate hypergeometric distribution (5.12) with the restriction $c \leq \sum_{i=1}^{s} x_i \leq d$ satisfies the Feller-Moran inequalities:

$$c\delta_i - 1 < x_i \leq (d+s)\delta_i, \quad i = 1,2,\ldots,s, \text{ where}$$

$$\delta_i = \frac{a_i+1}{\sum_{i=1}^{s} a_i + s} \text{ and } c \leq \sum_{i=1}^{s} x_i \leq d.$$

REFERENCES

[1] Feller, W. An Introduction to Probability Theory and Its Applications. New York: John Wiley and Sons, 1957.

[2] Finucan, H. M. "The Mode of a Multinomial Distribution." Biometrika, 51 (1964): 513-17.

[3] Janardan, K. G., and Patil, G. P. "A Unified Approach for a Class of Hypergeometric Models." Technical Report No. 16, 1969, Dept. of Statistics, The Pennsylvania State University.

[4] Patil, G. P. "On Sampling with Replacement from Population with Multiple Characters." Sankhya Series B 30 (1968): 355-66.

[5] Patil, G. P., and Joshi, S. W. A Dictionary and Bibliography of Discrete Distributions. Edinburgh: Oliver and Boyd, 1968; and New York: Hafner Publ. Co.

ELEMENTARY CHARACTERIZATION OF DISCRETE DISTRIBUTIONS

Z. GOVINDARAJULU
R. T. LESLIE
Department of Statistics
University of Kentucky

SUMMARY

If the conditional distribution of a set of discrete random variables for a specified value of the sum of the variables is given, then the distributions of the summands are characterized, provided certain conditions are realized. Further, there is a discussion of the minimum number of the determinations of the conditional distribution that are required in order to characterize the distributions of the summands. Several examples are also provided.

I. INTRODUCTION

Characterization of distributions has implications in statistical inference, and among discrete variables there are certain relationships of interest. One such relationship is that, if X and Y are independent Poisson variables, the conditional distribution of X is binomial for a given X+Y. Moran [3] and Chatterji [1] have shown that, if the conditional distribution of X (for a given X+Y, where X and Y are independent) is binomial, then X and Y are distributed as Poisson. Patil and Seshadri [6] proved a general characterization theorem which, in particular, includes Moran's result.

In all the above theorems independence of X and Y is a basic assumption rather than a deduction. Consequently, the aim of this paper is to explore and describe further the structure of the class of all conditional discrete distributions. Chatterji [1] has also given the number of the determinations of the conditional distribution of X (namely, the binomial distribution) for given X+Y, that are necessary in order to characterize the distributions of X and Y. As a matter of further interest, we shall also extend Chatterji's results to a set of c-variables.

2. MAIN RESULTS

In this section we derive the basic result which pertains to characterizing the structure of all conditional discrete distributions (for the bivariate case), and we provide several illustrations for the basic theorem.

Theorem 2.1: Let $\phi(t,k) > 0$ be defined for all t and $k(\leq t)$ non-negative integers; let $\sum_{k=0}^{t} \phi(t,k) = 1$ for all t; and let

$$P_k = \prod_{i=1}^{k} \frac{\phi(i,0)}{\phi(i,1)}, \quad Q_k = \prod_{i=1}^{k} \frac{\phi(i,i)}{\phi(i,i-1)}.$$

Then necessary and sufficient conditions that $\phi(t,k)$ should represent $P[X = k \mid X+Y = t]$, with X and Y independent, non-negative integer variates, are that

$$\chi(t,k) = \frac{\phi(t,k)}{\phi(t,k-1)} \quad \frac{\phi(t-1,k-1)}{\phi(t-1,k-2)}$$

should be independent of t ($2 \le k \le t$) and

$$\sum_{k=0}^{\infty} P_k \theta^k, \quad \sum_{k=0}^{\infty} Q_k \theta^k$$

should have positive radii of convergence.

Proof: Sufficiency: Let $\chi(t,k) = \chi(k)$, so that $\chi(t-i,k) = \chi(k)$ for $t-i \ge k$.

Then

$$\chi(k-i) = \frac{\phi(t-i,k-i)}{\phi(t-i,k-i-1)} \bigg/ \frac{\phi(t-i-1,k-i-1)}{\phi(t-i-1,k-i-2)}, \quad i = 0,1,\ldots,k-2,$$

and we may take the continued product of $\chi(i)$, for $i = 2,3,\ldots,k$ to obtain

$$\chi_1(k) = \prod_{i=2}^{k} \chi(k-i) = \frac{\phi(t,k)}{\phi(t,k-1)} \cdot \frac{\phi(t-k+1,0)}{\phi(t-k+1,1)}$$

$$= \frac{\phi(t,k)}{\phi(t,k-1)} \psi_1 (t-k+1), \tag{2.1}$$

where $\psi_1(k) = \frac{\phi(k,0)}{\phi(k,1)}$; putting $t = k = 1$ in (2.1) gives $\chi_1(1) = 1$.

We rewrite (2.1) as

$$\frac{\phi(t,k)}{\phi(t,k-1)} = \frac{\chi_1(k)}{\psi_1(t-k+1)}. \tag{2.2}$$

Taking the continued product of (2.2) with constant t and descending k, we have

$$\frac{\phi(t,k)}{\phi(t,0)} = \prod_{i=1}^{k} \frac{\phi(t,i)}{\phi(t,i-1)} = \prod_{i=1}^{k} \frac{\chi_1(i)}{\psi_1(t-i+1)} =$$

$$= \frac{\prod_{i=1}^{k} \chi_1(i) \prod_{i=1}^{t-k} \psi_1(i)}{\prod_{i=1}^{t} \psi_1(i)} = \frac{\chi_2(k)\psi_2(t-k)}{\psi_2(t)}, \tag{2.3}$$

where $\chi_2(k) = \prod_{i=1}^{k} \chi_1(i)$; $\psi_2(k) = \prod_{i=1}^{k} \psi_1(i)$; and $\chi_2(0) = \psi_2(0) = 1$.

Now $\sum_{k=0}^{t} \phi(t,k) = 1$, whence from (2.3)

$$\frac{1}{\phi(t,0)} = \frac{\sum_{k=0}^{t} \chi_2(k)\psi_2(t-k)}{\psi_2(t)} = \frac{\sum_{k=0}^{t} \chi_2(t-k)\psi_2(k)}{\psi_2(t)}.$$

Therefore $\phi(t,k) = \dfrac{\chi_2(k)\psi_2(t-k)}{\zeta_t}$, from (2.3),

with ζ_t the convolution of $\chi_2(k)$ and $\psi_2(k)$: $\prod_{k=0}^{t} \chi_2(k)\psi_2(t-k)$.

We now evaluate $\chi_2(k)$ and $\psi_2(k)$:

In (2.1) put k = t: $\chi_1(t) = \dfrac{\phi(t,t)}{\phi(t,t-1)} \psi_1(1)$. In particular,

$1 = \chi_1(1) = \dfrac{\phi(1,1)}{\phi(1,0)} \psi_1(1)$, so that $\psi_1(1) = \dfrac{\phi(1,0)}{\phi(1,1)}$.

$$\text{Therefore } \chi_2(k) = \prod_{i=1}^{k} \chi_1(i) = [\psi_1(1)]^k \prod_{i=1}^{k} \frac{\phi(i,i)}{\phi(i,i-1)} = Q_k\, P_1^k \tag{2.4}$$

$$\text{Also } \psi_2(k) = \prod_{i=1}^{k} \psi_1(i) = \prod_{i=1}^{k} \frac{\phi(i,0)}{\phi(i,1)} = P_k.$$

By hypothesis, there exists a $\underline{\theta}$ such that $\sum_{k=0}^{\infty} P_k\theta^k$ and $\sum_{k=0}^{\infty} Q_k\theta^k$ converge.

Let $P = \sum_{k=0}^{\infty} P_k\theta^k$ and $Q = \sum_{k=0}^{\infty} Q_k(P_1\theta)^k$; and write

$$f(k) = \frac{Q_k(P_1\theta)^k}{Q}, \quad g(k) = \frac{P_k\theta^k}{P}.$$

Thus f(k), g(k), k = 0,1,2,..., are frequency functions of variates X and Y, respectively.

$$\text{Moreover } \phi(t,k) = \frac{\chi_2(k)\psi_2(t-k)}{\zeta_t} = \frac{Q_k P_1^k P_{t-k}}{\zeta_t}$$

$$= \frac{Q_k(P_1\theta)^k/QP_{t-k}\theta^{t-k}/P}{\zeta_t\theta^t/PQ}$$

$$= \frac{f(k)g(t-k)}{\sum_{k=0}^{t} f(k)g(t-k)} = P[X = k \mid X+Y = t].$$

We have completed the demonstration of sufficiency.

<u>Necessity</u>.

If $\phi(t,k) = \dfrac{f(k)g(t-k)}{\zeta_t}$, substitution shows immediately that

$$\chi(t,k) = \frac{f(k)f(k-2)}{f^2(k-1)}$$, which is independent of t.

Taking the continued product, for i = 2,3,...,k, we find from (2.1):

$$\frac{f(k)}{f(k-1)} = \prod_{i=2}^{k} \chi(i)\frac{f(1)}{f(0)} = \chi_1(k)\frac{f(1)}{f(0)}. \tag{2.5}$$

Again taking the continued product for i = 1,2,...,k, we have

$$f(k) = \prod_{i=1}^{k} \chi_1(i)\left[\frac{f(1)}{f(0)}\right]^k f(0) = \chi_2(k)\left[\frac{f(1)}{f(0)}\right]^k f(0), \text{ from (2.3)}$$

$$= Q_k\left[\frac{P_1 f(1)}{f(0)}\right]^k f(0), \text{ from (2.4).}$$

Since $\sum_{k=0}^{\infty} f(k) = 1$, it is clear that $\sum_{k=0}^{\infty} Q_k\theta^k$ converges for positive θ.

This proves the necessity, as a similar argument holds for $\sum_{k=0}^{\infty} P_k\theta^k$.

The implication of the theorem is that, while f(k) and g(k) define ϕ(t-k) uniquely, ϕ(t,k), in turn, characterizes f(k) and g(k) to within an arbitrary constant, θ. More useful forms for f(k) and g(k) will now be derived.

Suppose that the required conditions on ϕ(t,k) are satisfied. Then $\frac{\phi(t,k)}{\phi(t,k-1)} = \frac{f(k)}{f(k-1)} \cdot \frac{g(t-k)}{g(t-k+1)} = \frac{\chi_1(k)}{\psi_1(t-k+1)}$

from (2.2), and using (2.5) we directly identify

$\theta\chi_1(k) = \frac{f(k)}{f(k-1)}$, and $\theta\psi_1(k) = \frac{g(k)}{g(k-1)}$, where θ is arbitrary.

$$\text{Hence } f(k) = f(0)\theta^k \prod_{i=1}^{k} \chi_1(i) \text{ and } g(k) = g(0)\theta^k \prod_{i=1}^{k} \psi_1(i), \tag{2.6}$$

where f(0) and g(0) are obtained by imposing the restrictions

$$1 = \sum_{k=0}^{\infty} f(k) = \sum_{k=0}^{\infty} g(k).$$

It is clear there are four starting points for generating distributions:

(1) given $\phi(t,k)$, to find $f(k)$, $g(k)$ and ζ_t;

(2) given $\chi_1(i)$ and $\psi_1(i)$, to find $\phi(t,k)$, $f(k)$, and $g(k)$;

(3) given $f(k)$ and $g(k)$, to find $\phi(t,k)$ and ζ_t; and

(4) given ζ_t, to find $\phi(t,k)$ and $f(k)$, $g(k)$.

Examples:

Method 1: Given $\phi(t,k)$

(a) Let $\phi(t,k) = \binom{t}{k} p^k (1-p)^{t-k}$

$$\chi(t,k) = \frac{k-1}{k} \text{ and is independent of } t\text{, as required.}$$

$$\frac{\phi(t,k)}{\phi(t,k-1)} = \frac{t-k+1}{k} \cdot \frac{p}{1-p}\text{, and we put } \chi_1(k) = \frac{1}{k}\cdot\frac{p}{1-p} \text{ and } \psi_1(k) = \frac{1}{k}.$$

From (2.6) we have $f(k) = f(0)\, \frac{\theta^k}{k!} \left(\frac{p}{1-p}\right)^k$

$$g(k) = g(0)\, \frac{\theta^k}{k!}$$

so that $f(0) = e^{-\frac{\theta p}{1-p}}$ and $g(0) = e^{-\theta}$,

and finally: $f(k) = e^{-\frac{\theta p}{1-p}} \left(\frac{\theta p}{1-p}\right)^k \frac{1}{k!}$

$$g(k) = e^{-\theta}\, \frac{\theta^k}{k!}$$

Now $\phi(t,0) = (1-p)^t$, so $\zeta_t = \frac{f(0)g(t)}{\phi(t,0)} = \frac{e^{-\frac{\theta p}{1-p}} \cdot e^{-\theta}\theta^t}{(1-p)^t t!}$

$$= \frac{e^{-\frac{\theta}{1-p}}\theta^t}{(1-p)^t t!}$$

(b) Laplacian distribution: Let $\phi(t,k) = e^{kt}\alpha(k)\beta(t)$,

where $\beta(t) = \left[\sum_{k=0}^{t} e^{kt}\alpha(k)\right]^{-1}$.

Then $\phi(t,0) = \alpha(0)\beta(t) = \alpha(0)/\left[\sum_{k=0}^{t} e^{kt}\alpha(k)\right]$.

Consider

$$\frac{\phi(t,k)}{\phi(t,k-1)} = e^t\, \frac{\alpha(k)}{\alpha(k-1)} = e^{t-k+1} \cdot e^{k-1} \cdot \frac{\alpha(k)}{\alpha(k-1)},$$

$$\text{so } \psi_1(i) = e^{-i} \text{ and } \chi_1(i) = \left[e^{i-1}\frac{\alpha(i)}{\alpha(i-1)}\right]^{-1}.$$

Then $g(k) = g(0)\theta^k e^{-k(k+1)/2}$, $[g(0)]^{-1} = \sum_{k=0}^{\infty} \theta^k e^{-k(k+1)/2}$;

and $f(k) = f(0)\theta^k e^{k(k-1)/2} \cdot \frac{\alpha(k)}{\alpha(0)}$, $f(0) = \alpha(0)\left[\sum_{k=0}^{\infty} \theta^k e^{k(k-1)/2}\alpha(k)\right]^{-1}$,

and $\zeta_t = \dfrac{g(0)f(0)\theta^t e^{-t(t+1)/2}}{\alpha(0)\beta(t)}$.

(c) Polya Distribution: Let

$$\phi(t,k) = \binom{\alpha+k-1}{k}\binom{\beta+t-k-1}{t-k}\Big/\binom{\alpha+\beta+t-1}{t}.$$

Then

$$\frac{\phi(t,k)}{\phi(t,k-1)} = \frac{\alpha+k-1}{k} \cdot \frac{t-k+1}{t-k+\beta},$$

so $\chi_1(i) = \dfrac{\alpha+i-1}{i}$ and $\psi_1(i) = i/(\beta-1+i)$.

Thus, $f(k) = f(0)\theta^k\binom{\alpha+k-1}{k}$ and $g(k) = g(0)\theta^k\binom{\beta+k-1}{k}$,

and $\zeta_t = f(0)g(0)\theta^t\binom{\alpha+\beta+t-1}{t}$. This leads us to the following theorem:

Theorem 2.3: Let X and Y be independent and generalized negative binomial variables with parameters $(1-\theta,\alpha)$ and $(1-\theta,\beta)$. Then the conditional distribution of X, given X+Y, is Polya and conversely.

(d) Hypergeometric Distribution:

$$\phi(t,k) = \binom{n_1}{k}\binom{n_2}{t-k}\Big/\binom{n_1+n_2}{t}; \quad \frac{\phi(t,k)}{\phi(t,k-1)} = \frac{n_1-k+1}{k} \cdot \frac{t-k+1}{n_2-t+k};$$

so set $\chi_1(i) = (n_1-i+1)/i$ and $\psi_1(i) = (n_2-i+1)/i$.

Hence $f(k) = f(0)\theta^k\binom{n_1}{k}$ and $g(k) = g(0)\theta^k\binom{n_2}{k}$, where

$f(0) = (1+\theta)^{-n_1}$ and $g(0) = (1+\theta)^{-n_2}$.

Method 2: Given $\chi_1(i)$ and $\psi_1(i)$:

(a) Let $\chi_1(i) = \psi_1(i) = 1/i^2$, so that

$$f(k) = f(0)\frac{\theta^k}{(k!)^2}, \quad g(k) = g(0)\frac{\theta^k}{(k!)^2}.$$

Now $\sum_{k=0}^{\infty} \frac{\theta^k}{(k!)^2} = I_o(2\sqrt{\theta})$, the zeroth order Bessel function,

whence $f(k) = g(k) = \frac{\Sigma\theta^k/(k!)^2}{I_o(2\sqrt{\theta})}$.

Moreover $[\phi(t,0)]^{-1} = \frac{\sum_{k=0}^{t} \chi_2(t-k)\psi_2(k)}{\psi_2(t)} = \sum_{k=0}^{t} \binom{t}{k}^2 = \binom{2t}{t}$.

Hence $\zeta_t = \frac{f(0)g(t)}{\phi(t,0)} = \frac{\theta^t}{(t!)^2} \binom{2t}{t} \left[\frac{1}{I_o 2\sqrt{\theta}}\right]^2$.

(b) Let $\chi_1(i) = a$ and $\psi_1(i) = b^{-1}$, so that

$f(k) = f(0)\theta^k a^k$, with $f(0) = 1-\theta a$; and $g(k) = g(0)\theta^k b^{-k}$,

with $g(0) = 1-(\theta/b)$, and

$\zeta_f = (1-\theta a)^{-1}(1-\theta/b)^{-1}(1-ab)^{-1}(\theta/b)^t\left[1-(ab)^{t+1}\right]$. In

particular, if $a = b^{-1}$, then

$\zeta_t = (1-\theta a)^{-2}(\theta a)^t\ (t+1)$,

which is the negative binomial distribution.

(c) Set $\chi_1(i) = (n_1-i+1)/i$, and $\psi_1(i) = i/(n_2-i+1)$.

Then $f(k) = f(0)\binom{n_1}{k}\theta^k$ and $g(k) = g(0)\binom{n_2}{k}\theta^k$, where

$f(0) = (1+\theta)^{-n_1}$ and $g(0) = (1+\theta)^{-n_2}$ and

$$\zeta_t = \binom{n_1+n_2}{t}\left(\frac{\theta}{1+\theta}\right)^t\left(\frac{1}{1+\theta}\right)^{n_1+n_2-t}.$$

<u>Method 3</u>: Given f(k) and g(k).

Let $f(k) = g(k) = a^k(1-a)$ for $0<a<1$.

Then $\theta\chi_1(i) = \theta\psi_1(i) = \frac{f(i)}{f(i-1)} = a$, giving

$\chi_2(k) = \psi_2(k) = \left(\frac{a}{\theta}\right)^k$; $[\phi(t,0)]^{-1} = t+1$ and

$\zeta_t = \frac{f(0)g(t)}{\phi(t,0)} = (1-a)^2 a^t(t+1)$. Thus

$$\phi(t,k) = \frac{\chi_2(k)\psi_2(t-k)}{\zeta_t} = \left(\frac{a}{\theta}\right)^t \frac{1}{(1-a)^2 a^t(t+1)} = \left[(1-a)^2(t+1)\theta^t\right]^{-1}.$$

But $\sum_{k=0}^{t} \phi(t,k) = 1$, giving $\theta^t(1-a)^2 = 1$, so that $\phi(t,k) = \frac{1}{t+1}$.

Method 4: Given ζ_t. This is perhaps the most fruitful of all four methods.

(a) Vandermonde's convolution:

From $\zeta_t = \theta^t \sum_{k=0}^{t} \binom{p+k}{k}\binom{q+t-k}{t-k} = \theta^t\binom{t+p+q+1}{t} = \sum_{k=0}^{t} f(k)g(t-k)$

and $\phi(t,k) = \frac{f(k)g(t-k)}{\zeta_t}$, we identify

$f(k) = f(0)\binom{p+k}{k}\theta^k$

$g(t-k) = g(0)\binom{q+t-k}{t-k}\theta^{t-k}$, or $g(k) = g(0)\binom{q+k}{k}\theta^k$.

Thus $f(0) = (1-\theta)^{p+1}$, $g(0) = (1-\theta)^{q+1}$ and

$f(k) = \binom{p+k}{k}(1-\theta)^{p+1}\theta^k$, $g(k) = \binom{q+k}{k}(1-\theta)^q\theta^k$. Thus if

$$\phi(t,k) = \frac{\binom{p+k}{k}\binom{q+t-k}{t-k}}{\binom{t+p+q+1}{t}}$$ is a Polya distribution, f(k) and g(k)

are negative binomial distributions, and conversely.

(b) $\sum_{k=0}^{t} \binom{2t-2k}{t-k}\binom{2k}{k}\frac{1}{2k+1} = \frac{2^4(t!)^2}{(2t+1)!}$ (see [2], p. 121.)

Let $f(k) = \binom{2k}{k}\frac{\theta^{2k}}{2k+1} f(0)$

$g(k) = \binom{2k}{k}\theta^{2k} g(0)$. Here, $f(0) = g(0) = (1-4\theta^2)^{-1/2}$

Thus $f(k) = \frac{\binom{2k}{k}\frac{\theta^{2k}}{2k+1}}{\sqrt{1-4\theta^2}}$, $g(k) = \frac{\binom{2k}{k}\theta^{2k}}{\sqrt{1-4\theta^2}}$, and

$$\phi(t,k) = \frac{\binom{2k}{k}\frac{1}{2k+1}\binom{2t-2k}{t-k}(2t+1)!}{2^4(t!)^2}$$

(c) Abel's generalization of the binomial theorem:

$$\frac{(x+y+t)^t}{x} = \sum_{k=0}^{t} \binom{t}{k}(x+k)^{k-1}(y+t-k)^{t-k}$$

$$= \sum_{k=0}^{t} \frac{(x+k)^{k-1}}{k!} \cdot \frac{(y+t-k)^{t-k}}{(t-k)!}$$

Let $f(k) = \frac{(x+k)^{k-1}}{k!} \theta^k f(0)$ and $g(k) = \frac{(y+k)^k}{k!} \theta^k g(0)$,

with $[f(0)]^{-1} = \sum_{k=1}^{\infty} \frac{(x+k)^{k-1}}{k!} \theta^k$ and $[g(0)]^{-1} = \sum_{k=0}^{\infty} \frac{(y+1)^k}{k!} \theta^k$;

$$\phi(t,k) = \frac{\binom{t}{k}(x+k)^{k-1} (y+t-k)^{t-k} x}{(x+y+t)^t}$$

Remark: It is possible to extend Theorem 2.1 to the c-variate case; this is demonstrated in Theorem 3.3 below.

3. C-VARIATE EXTENSIONS.

In this section we shall state and prove three theorems which pertain to c-variate situations.

It is well known that, if the set of independent random variables $X_1,\ldots,X_c$ are each binomially distributed with the same parameter p for the probability of a success, then the conditional distribution of the X's, given $\sum_1^c X_i = t$, is generalized hypergeometric. Also, it is known that, if the X's are independent Poisson random variables, then the conditional distribution of the X's, given $\sum_1^c X_i = t$, is multinomial. In the following we shall state and prove the converses of these theorems, assuming the minimum number of determinations of the conditional distributions. Recall that [1] discusses the minimum number of determinations of the conditional distribution for the bivariate Poisson case. So, Theorem 3.2 constitutes a direct extension of Theorems 1 and 4 in [1] to c-variate situation. In Theorem 3.3, we shall extend Theorem 3.1 to more than 2 variates.

Theorem 3.1: Let $X_1,X_2,\ldots,$ be a sequence of independent random variables such that $P(X_i = j) = f_i(j) \geq 0$, $j = 0,1,\ldots,$ and $i = 1,2,\ldots$. Also, let

$$P\left\{X_k=t,X_{k+1}=0,X_i=0,[i=1,\ldots,c,(i\neq k,k+1)\,\Big|\,\sum_1^c X_i=t]\right\} = \binom{n_k}{t}\Big/\binom{\sum_{k=1}^{c} n_k}{t}$$

and

$$P\left[X_k=t-1,X_{k+1}=1,X_i=0,i=1,\ldots,c,(i\neq k,k+1)\,\Big|\,\sum_1^c X_i=t\right] =$$

$$= \binom{n_k}{t-1}\binom{n_{k+1}}{1} \Big/ \binom{\sum_1^c n_k}{t}, \qquad (3.1)$$

for $k = 1,\ldots,c-1$ and $t \geq 1$. Then

$$f_k(j) = \binom{n_k}{j} p^j (1-p)^{n_k - j}, \quad \frac{p}{1-p} = f_k(1)/n_k f_k(0),$$

$$j = 0,1,\ldots,n_k \text{ and } k = 1,\ldots,c.$$

Proof: For simplicity, let $k = 1$. Then, from the set of equations in (3.1) we obtain

$$f_1(t) = \frac{f_2(1)}{n_2 f_2(0)} \cdot \frac{n_1-t+1}{t} f_1(t-1) = \theta_1 \frac{n_1-t+1}{t} \cdot f_1(t-1),$$

where $\theta_1 = f_2(1)/n_2 f_2(0)$. Thus, taking repeated products, we have

$$f_1(t) = \binom{n_1}{t} \theta^t f_1(0),$$

where $[f_1(0)]^{-1} = \sum_{t=0}^{\infty} \binom{n_1}{t} \theta^t = (1+\theta)^{n_1}$

so, $f_1(t) = \binom{n_1}{t}\left(\frac{\theta}{1+\theta}\right)^t \cdot \left(\frac{1}{1+\theta}\right)^{n_1-t}$, $t = 0,1,\ldots,n_1$.

Analogously, one can establish that

$$f_k(t) = \binom{n_k}{t}\left(\frac{\theta_k}{1+\theta_k}\right)^t \cdot \left(\frac{1}{1+\theta_k}\right)^{n_k-t}, \quad \theta_k = \frac{f_{k+1}(1)}{n_{k+1} f_{k+1}(0)},$$

for $k = 1,2,\ldots,c-1$.

However, from the first equation in the hypothesis, we have

$$\frac{f_1(t) \prod_{j=2}^{c} f_j(0)}{\binom{n_1}{t}} = \frac{\zeta_t}{\binom{\sum_1^c n_k}{t}}, \text{ where } \zeta_t = P\left(\sum_1^c X_k = t\right),$$

$$\zeta_t = \binom{\sum_1^c n_k}{t} \cdot \theta_1^t \prod_{j=1}^{c} f_j(0).$$

Similarly we can show that $\zeta_t = \binom{\sum_1^c n_k}{t} \theta_k^t \prod_1^c f_j(0)$, $k = 2,\ldots,c-1$.

Hence $\theta_1 = \theta_2 = \cdots = \theta_{c-1} = \theta$. Now let $F_j(u) = \sum_{t=0}^{\infty} f_j(t) u^t$;

then

$$\prod_{j=1}^{c} F_j(u) = \prod_{j=0}^{c} \sum_{t=0}^{\infty} \binom{\sum_1^c n_k}{t} \theta^t u^t = (1+\theta u)^{\Sigma n_k}.$$

Since $F_j(u) = (1+\theta u)^{n_j}$, $j = 1,\ldots,c$, we have

$$F_c(u) = (1+\theta u)^{\Sigma n_k}/(1+\theta u)^{\sum_1^{c-1} n_j} = (1+\theta u)^{n_c}.$$

That is, $f_c(t) = \binom{n_c}{t} f_c(0)\, \theta^t$, $f_c(0) = (1+\theta)^{n_c}$.

This completes the proof of Theorem 3.1.

The following theorem constitutes an extension of Chatterji's Theorems 1 and 4 to the c-variate case.

Theorem 3.2: Let $X_1, X_2, \ldots,$ be a sequence of independent random variables with $P(X_i=j) = f_i(j) > 0$, $j = 0,1,\ldots,$ and $i = 1,2,\ldots$. Also, let

$$P\left[X_k = t, X_{k+1}=0, X_i=0 \text{ for } i=1,\ldots,c,\ (i\neq k,k+1) \,\middle|\, \sum_1^c X_i=t\right] = p_{k,t}^t;$$

$$P\left[X_k=t-1, X_{k+1}=1, X_i=0 \text{ for } i=1,\ldots,c,\ (i\neq k,k+1) \,\middle|\, \sum_1^c X_i=t\right]$$

$$= t p_{k,t}^{t-1} p_{k+1,t};$$

and

$$P\left[X_k=t-2, X_{k+2}=2, X_i=0, \text{for } i=1,\ldots,c,\ (i\neq k,k+1) \,\middle|\, \sum_1^c X_i=t\right]$$

$$= \frac{t!}{(t-2)!2!} \cdot p_{k,t}^{t-2} p_{k+1,t}^2, \tag{3.2}$$

for $k = 1,\ldots,c-1$ and $t \geq 2$. Then

$$p_{i,t} = p_i,\ i = 1,\ldots,c,\quad t = 0,1,\ldots, \text{ and}$$

$f_i(j) = e^{-\psi p_i}(\psi p_i)^j/j!$, where $\psi \equiv \theta_i \alpha_i/p_i$, $\alpha_i = p_i/p_{i+1}$,

and $\theta_i = f_{i+1}(1)/f_{i+1}(0)$, $i = 1,\ldots,c$.

Proof: For simplicity, let $k = 1$. Then, from the set of three equations (3.2) and using the independence of $X_1,\ldots,X_c$, we obtain:

$$\frac{f_1(t)f_2(0)}{f_1(t-1)f_2(1)} = \left(\frac{1}{t}\right) \frac{p_{1t}}{p_{2t}} \tag{3.3}$$

and

$$\frac{f_1(t-1)f_2(1)}{f_1(t-2)f_2(2)} = \frac{2}{t-1}\,\frac{p_{1t}}{p_{2t}} \,. \tag{3.4}$$

In other words,

$$f_1(t) = \theta_1 \frac{\alpha_{1t}}{t} f_1(t-1),\ \alpha_{1t} = p_{1t}/p_{2t}, \tag{3.5}$$

and

$$f_1(t-1) = \phi_1 \frac{\alpha_{1t}}{t-1} f_1(t-2),\ \phi_1 = 2f_2(2)/f_2(1) \tag{3.6}$$

From (3.5) and (3.6) we obtain

$$\alpha_{1t} = (\theta_1/\phi_1)\alpha_{1,t-1},\ \text{for } t \geq 2. \tag{3.7}$$

Hence

$$\alpha_{1t} = (\theta_1/\phi_1)^{t-1}\alpha_{1,1}$$

and

$$f_1(t) = \frac{\theta_1^t}{t!} \prod_{j=1}^{t} \alpha_{ij}\ f_1(0)$$

$$= \frac{(\theta_1\alpha_{11})^t}{t!} \left(\frac{\theta_1}{\phi_1}\right)^{t(t-1)/2} f_1(0). \tag{3.8}$$

Also since

$$\sum_{t=1}^{\infty} f_1(t) \text{ converges, } \theta_1/\phi_1 \leq 1.$$

Now, we will show that $\theta_1/\phi_1 \geq 1$.

From equation (3.2) we have

$$f_1(t)\ Q_{c-1}/p_{1t}^t = \zeta_t,$$

where $\zeta_t = P \sum_1^c x_i = t$ and $Q_{c-i+1} = f_i(0),\ldots,f_c(0)$, $i = 1,\ldots,c$.

Thus

$$Q_{c-1}\left[f_1(0) + \sum_{t=1}^{\infty} \frac{f_1(t)\ u^t}{p_{1t}^t}\right] = \prod_{j=1}^{c} F_j(u),$$

where $F_j(u) = \sum_{t=0}^{\infty} f_j(t)u^t$. Now, using (3.8), we have

$$\prod_1^c F_j(u) = Q_{c-1}\left[f_1(0) + f_1(0)\sum_{t=1}^{\infty} \cdot \frac{1}{t!}\left(\frac{\theta_1\alpha_{11}}{p_{1t}}\right)^t\left(\frac{\theta_1}{\phi_1}\right)^{t(t-1)/2} u^t\right]. \tag{3.9}$$

However $\frac{\alpha_{11}}{p_{1t}} \geq \alpha_{11}\left(\frac{p_{1t}+p_{2t}}{p_{1t}}\right) = \alpha_{11}\left(1 + \frac{1}{\alpha_{1t}}\right) = \alpha_{11} + \left(\frac{\phi_1}{\theta_1}\right)^{t-1}$;

consequently,

$$\prod_1^c F_j(u) \geq Q_c\left\{1 + \sum_{t=1}^{\infty} \frac{1}{t!}\,\theta_1^t\left(\frac{\theta_1}{\phi_1}\right)^{t(t-1)/2}\left[\alpha_{11} + \left(\frac{\phi_1}{\theta_1}\right)^{t-1}\right]^t u^t\right\}$$

$$= Q_c\left(1 + \sum_{t=1}^{\infty} d_t\, u^t\right) \quad \text{(say)}.$$

Consider

$$\frac{d_{t+1}}{d_t} = \frac{\theta_1}{t+1}\left(\frac{\theta_1}{\phi_1}\right)^t\left[\frac{\alpha_{11}+\frac{\phi_1}{\theta_1}^{t}}{\alpha_{11}+\frac{\phi_1}{\theta_1}^{t-1}}\right]^t\left[\alpha_{11}+\left(\frac{\phi_1}{\theta_1}\right)^t\right].$$

If $\theta_1 < \phi_1$, $\frac{d_{t+1}}{d_t} \sim \frac{\theta_1}{t+1}\left(\frac{\phi_1}{\theta_1}\right)^t > \theta_1 \frac{1}{t+1}$,

which implies that the series $\Sigma d_t\, u^t$ diverges, and this leads to a contradiction, since the lefthand side is $\prod_1^c F_j(u)$. Thus we infer that $\phi_1 \leq \theta_1$. Since we have already established that $\theta_1 \leq \phi_1$, we obtain $\theta_1 = \phi_1$, and hence $\alpha_{1t} \equiv \alpha_1$. As a result, (3.8) becomes

$$f_1(t) = f_1(0)\,(\theta_1\alpha_1)^t/t!,\quad f_1(0) = e^{-\theta_1\alpha_1} \tag{3.10}$$

and hence $F_1(u) = \sum_{t=0}^{\infty} f_1(t)u^t = e^{\theta_1\alpha_1(u-1)}$. (3.11)

Analogously one can show that

$$f_k(t) = e^{-\theta_k\alpha_k}\,(\theta_k\alpha_k)^t/t!,\quad t = 0,1,\ldots,$$

and $k = 1,2,\ldots,c-1$. (3.12)

Now, setting $\theta_1 = \phi_1$ in (3.9), we obtain

$$\prod_1^c F_j(u) = Q_c\left[1 + \sum_{t=1}^{\infty}\left(\frac{\theta_1\alpha_1}{p_1}\right)^t u^t\right]. \tag{3.13}$$

In general, one can establish that

$$\prod_{1}^{c} F_j(u) = Q_c\left[1 + \sum_{i=1}^{\infty}\left(\frac{\theta_k\alpha_k}{p_k}\right)^t u^t\right]; \tag{3.14}$$

that is,

$$\frac{\theta_k\alpha_k}{p_k} \equiv \psi \quad \text{(a constant)}. \tag{3.15}$$

Hence

$$f_k(t) = e^{-\psi p_i}(\psi p_i)^t/t!, \; t = 0,1,\ldots, \tag{3.16}$$

and

$$F_k(u) = e^{(u-1)\psi p_k}, \; k = 1,\ldots,c-1. \tag{3.17}$$

So,

$$F_c(u) = e^{\psi u - \sum_{k=1}^{c-1} \psi u p_k} = f_c(0)e^{\psi u p_c}. \tag{3.18}$$

Hence

$$f_c(t) = e^{-\psi p_c}(\psi p_c)^t/t!, \; t = 0,1,\ldots. \tag{3.19}$$

This completes the proof of Theorem 3.2.

Extension of Theorem 2.1 to the c-variate case is non-trivial, the main problem being to find the additional conditions on $\phi(t,k_1,k_2,\ldots,k_{c-1})$ to ensure independence of the unconditional distributions. For simplicity we state and prove the theorem for c = 3 and allow this to indicate the generalization to higher dimensions.

<u>Theorem 3.3</u>: Let $\phi(t,k_1,k_2) >$ be defined for all t, $k_1,k_2(k_1+k_2 \le t)$ non-negative integers, let $\sum_{k_1=0}^{t}\sum_{k_2=0}^{t}\phi(t,k_1,k_2) = 1$ for all t, and let

$$Q_{k_1} = \prod_{i=1}^{k_1}\frac{\phi(i,i,0)}{\phi(i,i-1,0)}, \qquad Q_{k_2} = \prod_{i=1}^{k_2}\frac{\phi(i,0,i)}{\phi(i,0,i-1)}, \text{ and}$$

$$P_i = \prod_{j=1}^{i}\frac{\phi(j,0,0)}{\phi(j,0,1)} = \prod_{j=1}^{i}\frac{\phi(j,0,0)}{\phi(j,1,0)}\,\eta, \; \eta \text{ a constant.}$$

Then necessary and sufficient conditions that $\phi(t,k_1,k_2)$ ahould represent $P[X_1=X_1,X_2=k_2|X_1+X_2+X_3=t]$, with X_1,X_2,X_3

independent non-negative integer variates, are that

$$\chi^{(1)}(t,k_1,k_2) = \frac{\phi(t,k_1,k_2)\phi(t-1,k_1-2,k_2)}{\phi(t,k_1-1,k_2)\phi(t-1,k_1-1,k_2)} \quad (2 \le k_1 \le t-k_2)$$

be a function of k_1 alone;

$$\chi^{(2)}(t,k_1,k_2) = \frac{\phi(t,k_1,k_2)\phi(t-1,k_1,k_2-2)}{\phi(t,k_1,k_2-1)\phi(t-1,k_1,k_2-1)} \quad (2 \le k_2 \le t-k_1)$$

be a function of k_2 alone;

$$\phi(t,k_1,k_2)\phi(t-1,k_1-1,k_2-1) = \phi(t,k_1,k_2-1)\phi(t-1,k_1-1,k_2)$$

$$= \phi(t,k_1-1,k_2)\phi(t-1,k_1,k_2-1),$$

and $\sum_{k_1=0}^{\infty} Q_{k_1}\theta^{k_1}$, $\sum_{k_2=0}^{\infty} Q_{k_2}\theta^{k_2}$, $\sum_{i=0}^{\infty} P_i\theta^i$ should have positive radii of convergence.

Proof: Sufficiency: Let $\chi^{(1)}(t,k_1,k_2) = \chi^{(1)}(k_1)$ and $\chi^{(2)}(t,k_1,k_2) = \chi^{(2)}(k_2)$, so that

$$\chi^{(1)}(t-i,k_1,k_2-i) = \chi^{(1)}(k_1) \text{ and } \chi^{(2)}(t-i,k_1-i,k_2)$$

$$= \chi^{(2)}(k_2), \text{ for } t-i \ge k_1 \text{ and } t-i \ge k_2, \text{ respectively.}$$

Then $\chi^{(1)}(k_1-i) = \dfrac{\phi(t-i,k_1-i,k_2)}{\phi(t-i,k_1-i-1,k_2)} \cdot \dfrac{\phi(t-i-1,k_1-i-2,k_2)}{\phi(t-i-1,k_1-i-1,k_2)}$,

for $i = 0,1,\ldots,k_1-2$;

$$\chi^{(2)}(k_2-i) = \frac{\phi(t-i,k_1,k_2-i)}{\phi(t-i,k_1,k_2-i-1)} \cdot \frac{\phi(t-i-1,k_1,k_2-i-2)}{\phi(t-i-1,k_1,k_2-i-1)}$$

for $i = 0,1,\ldots,k_2-2$;

and we may take the continued product of $\chi^{(1)}(i)$ for $i = 2,3,\ldots,k_1$, $\chi^{(2)}(i)$ for $i = 2,3,\ldots,k_2$ to obtain

$$\chi_1^{(1)}(k_1) = \prod_{i=0}^{k_1-2} \chi^{(1)}(k_1-i) = \frac{\phi(t,k_1,k_2)}{\phi(t,k_1-1,k_2)} \cdot \frac{\phi(t-k_1+1,0,k_2)}{\phi(t-k_1+1,1,k_2)},$$

a function of $k_1(>1)$ alone (3.20)

$$\chi_1^{(1)}(1) = 1$$

and

$$\chi_1^{(2)}(k_2) = \prod_{i=0}^{k_2-2} \chi^{(2)}(k_2-i) = \frac{\phi(t,k_1,k_2)}{\phi(t,k_1,k_2-1)} \cdot \frac{\phi(t-k_2+1,k_1,0)}{\phi(t-k_2+1,k_1,1)},$$

a function of k_2(1) alone (3.21)

$$\chi_1^{(2)}(1) = 1.$$

From $\frac{\phi(t,k_1,k_2)}{\phi(t,k_1-1,k_2)} = \frac{\phi(t-1,k_1,k_2-1)}{\phi(t-1,k_1-1,k_2-1)}$ we have

$$\frac{\phi(t-k_1+1,0,k_2)}{\phi(t-k_1+1,1,k_2)} = \frac{\phi(t-k_1,0,k_2-1)}{\phi(t-k_1,1,k_2-1)} = \cdots = \frac{\phi(t-k_1-k_2+1,0,0)}{\phi(t-k_1-k_2+1,1,0)};$$

and $\frac{\phi(t,k_1,k_2)}{\phi(t,k_1-1,k_2)} = \frac{\phi(t-k_2,k_1,0)}{\phi(t-k_2,k_1-1,0)}$; and from

$\frac{\phi(t,k_1,k_2)}{\phi(t,k_1,k_2-1)} = \frac{\phi(t-1,k_1-1,k_2)}{\phi(t-1,k_1-1,k_2-1)}$ we have

$$\frac{\phi(t-k_2+1,k_1,0)}{\phi(t-k_2+1,k_1,1)} = \frac{\phi(t-k_2,k_1-1,0)}{\phi(t-k_2,k_1-1,1)} = \cdots = \frac{\phi(t-k_1-k_2+1,0,0)}{\phi(t-k_1-k_2+1,0,1)},$$

and $\frac{\phi(t,k_1,k_2)}{\phi(t,k_1-1,k_2)} = \frac{\phi(t-k_1,0,k_2)}{\phi(t-k_1,0,k_2-1)}$.

Therefore, $$\chi_1^{(1)}(k_1) = \frac{\phi(t,k_1,k_2)}{\phi(t,k_1-1,k_2)} \cdot \frac{\phi(t-k_1-k_2+1,0,0)}{\phi(t-k_1-k_2+1,1,0)} \tag{3.22}$$

and $$\chi_1^{(1)}(k_2) = \frac{\phi(t,k_1,k_2)\phi(t-k_1-k_2+1,0,0)}{\phi(t,k_1,k_2-1)\phi(t-k_1-k_2+1,0,1)} . \tag{3.23}$$

Again taking continued products of (3.22) and (3.23) for fixed t and reducing k_1 and k_2, respectively, we find

$$\chi_2^{(1)}(k_1) = \prod_{i=1}^{k_1} \chi_1^{(1)}(k_1-i) = \frac{\phi(t,k_1,k_2)}{\phi(t,0,k_2)} \cdot \prod_{i=1}^{k_1} \frac{\phi(t-i-k_2+1,0,0)}{\phi(t-i-k_2+1,1,0)} \tag{3.24}$$

and

$$\chi_2^{(2)}(k_2) = \prod_{i=1}^{k_2} \chi_1^{(2)}(k_2-i) = \frac{\phi(t,k_1,k_2)}{\phi(t,k_1,0)} \prod_{i=1}^{k_2} \frac{\phi(t-k_1-i+1,0,0)}{\phi(t-k_1-i+1,0,1)}. \tag{3.25}$$

Hence $\phi(t,k_1,k_2) = \chi_2^{(1)}(k_1) \frac{\psi_1(t-k_2)}{\psi_1(t-k_1-k_2)} \phi(t,0,k_2)$, from (3.24),

where $\psi_1(i) = \prod_{j=1}^{i} \frac{\phi(j,0,0)}{\phi(j,1,0)}$; and

$$\phi(t,k_1,k_2) = \chi_2^{(2)}(k_2)\frac{\psi_2(t-k_1)}{\psi_2(t-k_1-k_2)}\phi(t,k_1,0),\text{ from (3.25), where}$$

$$\psi_2(i) = \prod_{j=1}^{i}\frac{\phi(j,0,0)}{\phi(j,0,1)}\ .$$

Noting that $\psi_1(t-k_2)\phi(t,0,k_2)$ is independent of k_1, while $\psi_2(t-k_1)\phi(t,k_1,0)$ is independent of k_2, the former must be of the form $\alpha(t)\chi_2^{(2)}(k_2)$ and the latter, $\alpha(t)\chi_2^{(1)}(k_1)$, in order that the two forms of $\phi(t,k_1,k_2)$ be identical for all t,k_1,k_2.

$$\text{Thus } \phi(t,k_1,k_2) = \frac{\chi_2^{(1)}(k_1)\chi_2^{(2)}(k_2)\chi_2^{(3)}(t-k_1-k_2)}{\chi_2^{(4)}(t)}\ ,$$

where $\psi_1(t-k_1-k_2) = \psi_2(t-k_1-k_2)$, implying that $\phi(i,1,0) = \phi(i,0,1)$, $i \geq 1$; and with suitable normalization, we may write

$$\phi(t,k_1,k_2) = \frac{f(k_1)g(k_2)h(t-k_1-k_2)}{\zeta_t} = \Pr[X_1=k_1,X_2=k_2|X_1+X_2+X_3=t],$$

where ζ_t is the three-fold convolution of f,g,h.

We now evaluate $\chi_2^{(1)}(k_1)$, $\chi_2^{(2)}(k_2)$, and $\chi_2^{(3)}(t-k_1-k_2)$. In (3.20) put $k_1 = t$ and $k_2 = 0$: $\chi_1^{(1)}(t) = \frac{\phi(t,t,0)}{\phi(t,t-1,0)}P_1$. In particular, $\chi_1^{(1)}(1) = 1$. In (3.21) put $k_1 = 0$ and $k_2 = t$: $\chi_1^{(2)}(t) = \frac{\phi(t,0,t)}{\phi(t,0,t-1)}P_1$. In particular, $\chi_1^{(2)}(1) = 1$.

$$\text{Thus, } \chi_2^{(1)}(k_1) = \prod_{i=1}^{k_1}\chi_1^{(1)}(i) = P_1^{k_1}\prod_{i=1}^{k_1}\frac{\phi(i,i,0)}{\phi(i,i-1,0)} = P_1^{k_1}Q_{k_1},$$

$$\text{and } \chi_2^{(2)}(k_2) = \prod_{i=1}^{k_2}\chi_1^{(2)}(i) = P_1^{k_2}\prod_{i=1}^{k_2}\frac{\phi(i,0,i)}{\phi(i,0,i-1)} = P_1^{k_2}Q_{k_2}.$$

$$\text{Moreover, as noted above, } \chi_2^{(3)}(i) = \prod_{j=1}^{i}\frac{\phi(j,0,0)}{\phi(j,1,0)} = P_i,$$

By hypothesis, there exists a θ such that $\sum_{k_1=0}^{\infty} Q_{k_1}\theta^{k_1}$, $\sum_{k_2=0}^{\infty} Q_{k_2}\theta^{k_2}$, and $\sum_{k_3=0}^{\infty} P_{k_3}\theta^{k_3}$ converge.

Let $Q_1 = \sum_{k_1=0}^{\infty} Q_{k_1}(P_1\theta)^{k_1}$, $Q_2 = \sum_{k_2=0}^{\infty} Q_{k_2}(P_1\theta)^{k_2}$, and

$Q_3 = \sum_{k_3=0}^{\infty} P_{k_3}\theta^{k_3}$; and write $f(k_1) = \dfrac{Q_{k_1}(P_1\theta)^{k_1}}{Q_1}$,

$g(k_2) = \dfrac{Q_{k_2}(P_1\theta)^{k_2}}{Q_2}$, $h(k_3) = \dfrac{P_{k_3}\theta^{k_3}}{Q_3}$. $f(k)$, $g(k)$, and $h(k)$ are

thus frequency functions, of variates X_1, X_2, X_3, say.

$$\text{Moreover, } \phi(t,k_1,k_2) = \frac{\chi_2^{(1)}(k_1)\chi_2^{(2)}(k_2)\chi_3^{(3)}(t-k_1-k_2)}{\zeta_t}$$

$$= \frac{Q_{k_1}(P_1\theta)^{k_1} Q_{k_2}(P_1\theta)^{k_2} P_{t-k_1-k_2}\theta^{t-k_1-k_2}}{Q_1Q_2Q_3 \cdot \zeta_t\theta^t/Q_1Q_2Q_3}$$

$$= \frac{f(k_1)g(k_2)h(t-k_1-k_2)}{\sum_{k_1=0}^{t}\sum_{k_2=0}^{t} f(k_1)g(k_2)h(t-k_1-k_2)} .$$

Necessity follows along the same lines as in Theorem 2.1.

Now the c-variate version of Theorem 3.3 and its proof can be formulated. For the sake of simplicity, they are suppressed here.

Application:

Assume that there are k candidates for election to a certain office. Take random samples of size t and determine the conditional distribution of

$$X_1,\ldots,X_k, \text{ given } \Sigma X_i = t;$$

then assuming a certain form for $P(X_i=k_i, i = 1,\ldots,c \mid \sum_1^c X_i = t)$, we can determine the unconditional distributions of X_i.

Concluding remark: The results of Sections 2 and 3 are valid for all continuous distributions admitting density functions. The necessary modifications to be made are as follows: the probability function should be interpreted throughout as the density function and summations must be replaced by the appropriate integrals.

REFERENCES

[1] Chatterji, S. D. 1963. Some elementary characterizations of the Poisson distribution. American Mathematics Monthly 70, 9:958-64.

[2] Riordan, J. 1968. Combinatorial Identities. John Wiley & Sons, New York.

[3] Moran, P. A. P. 1951. A characteristic property of the Poisson distribution. Proceedings of Cambridge Philosophical Society 48:206-7.

[4] National Bureau of Standards. 1968. Handbook of Mathematical Functions. Applied Mathematics Series No. 55.

[5] Patil, G. P. 1965. On a characterization of multi-variate distribution by a set of its conditional distributions. Paper presented at the 35th session of the International Statistical Institute (held at Belgrade, Yugoslavia).

[6] ———, and Seshadri, V. 1964. Characterization theorems for some univariate probability distributions. The Journal of the Royal Statistical Society Series B 26:286-92.

APPROXIMATIONS TO THE GENERALIZED NEGATIVE BINOMIAL DISTRIBUTION*

HAJIME MAKABE
Tokyo Institute of Technology

Z. GOVINDARAJULU
University of Kentucky

SUMMARY

The generalized negative binomial distribution was introduced by one of the authors, and its asymptotic normality was discussed under suitable conditions. However, it is of interest to improve the normal approximation and to explore other approximations. In this paper a correction term to the normal approximation with the error term suitably evaluated, is derived. Also, approximations by Poisson distribution and negative binomial distribution with the correction terms defined by difference of itself are obtained. Further, the error term in the approximation to the distribution of a sum of independent random variables, which are rarely different from zero, by infinitely divisible law is evaluated, and some considerations are described. The main tools are expansion of the characteristic function and application of Lévy's inversion formula.

* This research was supported in part by the National Science Foundation Grant NSF GP 5664.

1. INTRODUCTION

The random variable $S_n = X_1+X_2+\ldots+X_n$, where the X's are independent (not necessarily identical) geometric random variables with probabilities

$$P(X_k=j) = p_k^j q_k, \quad j = 0,1,2,\ldots, \quad k = 1,2,\ldots,n, \tag{1.1}$$

is called a generalized negative binomial random variable. Govindarajulu [2] has defined the distribution of S_n and obtained a sufficient condition for the asymptotic normality of S_n, when suitably standardized. Considerable emphasis is placed on obtaining Poisson and negative binomial approximations for the same purpose. Makabe [9] has derived Poisson and binomial approximations to the generalized positive binomial (Poisson-binomial) distribution with a bound for the error term. However, there is no analogous relation between generalized positive and negative binomial distributions such as we find between proper positive and negative binomials (see [10]).

First, we shall consider the normal approximation to S_n under the restriction $p_i < 0.17$ $(i = 1,2,\ldots,n)$, although it is possible to weaken this restriction to $p_i < 1/3$ $(i = 1,2,\ldots,n)$. In order to obtain the refined boundary for the error term, the restriction $p_i < 0.17$ $(i = 1,2,\ldots,n)$ has to be imposed. This result is given in Theorem 1, and its proof is similar to that in [6]. Next, in Theorem 2, assuming that the p_i are small, a Poisson approximation with correction term defined by the second order difference of cumulative Poisson, is developed. (For generalized positive binomial distribution, similar results were obtained in [4], [8], [1], and [3].) Further, it is shown that the negative binomial approximation is also appropriate, with the boundary on the error term of the same order of magnitude as in the case of Poisson approximation. This relation, except for the exact boundary for the error obtained, is also evident from Theorem 1 of [10], which implies that the distance of the distributions defined by two sums of independent random variables S_n and S_n', is of the order $E(|S_n-S_n'|^3)$, when $ES_n = ES_n'$ and $ES_n^2 = ES_n'^2$.

Third, LeCam's [5] result concerning approximation by infinitely divisible law to that of the sum of independent random variables (which are rarely different from zero) is improved in discrete cases (see Theorem 4); and the order $\alpha^{1/3}$ of the boundary for the error term is replaced by α, where $\alpha = \max_{1\le i\le n} p_i$, and $p_i = P(X_i \neq 0)$. Poisson approximation to the same is obtained in [1] and [3] with the boundary for the error term expressed in terms of the first and second moments. The approximation of Theorem 4 in the special case of the generalized negative binomial distribution is compared with the other approximations obtained in this paper.

2. NORMAL APPROXIMATION

Govindarajulu [2] proved the following theorem:

Theorem 2.1: If

$$\sum_{k=1}^{n}\left[p_k(1+3P_k)q_k^{-3}\right] \Big/ \left[\sum_{k=1}^{n} p_k q_k^{-2}\right]^{3/2} \to 0 \text{ as } n\to\infty$$

then, uniformly in ℓ we have

$$\lim_{n\to\infty} \sum_{j=0}^{\ell} P(S_n=j) = (2\pi)^{-1/2}\int_{-\infty}^{x} e^{-u^2/2}du, \text{ where}$$

$$x = \left(\ell - \sum_{k=1}^{n} p_k q_k^{-1}\right) \Big/ \left(\Sigma p_k q_k^{-2}\right)^{1/2}.$$

NOTE: It was also pointed out by [2] that $q_k > 2/3$, $(k = 1,2,\ldots,n)$ is also sufficient for the asymptotic normality of S_n.

In the following theorem a correction term to the above approximation, together with a boundary for the error term, would be obtained.

Theorem 2.2: If $p_k \le 0.17$ for all k, and $\sigma^2 = \sum_{k=1}^{n} p_k q_k^{-2} > 25$, then we have

$$\sum_{j=\ell+1}^{m} P(S_n=j) = \int_{\xi_1}^{\xi_2} \frac{1}{\sqrt{2\pi}} e^{-\frac{u^2}{2}}du + \frac{A}{\sqrt{2\pi\sigma}} \quad (1-\xi_2^2)\, e^{-\frac{\xi_2^2}{2}} - (1-\xi_1^2)\, e^{-\frac{\xi_1^2}{2}} + \P$$

$$\text{where } A = \sum_{k=1}^{n} \frac{1+p_k}{6q_k}\left(\frac{p_k}{q_k^2}\right) \text{ and}$$

$$|\P| \le \frac{0.17}{\sigma^2} + \frac{0.51}{\sigma^3} + \frac{0.99}{\sigma^4} + \frac{0.13}{\sigma^5} + \frac{1.09}{\sigma^7} + \frac{0.31}{\sigma^9} + \frac{2}{3\sigma}\left(e^{-\frac{3}{4}\sigma} + e^{-\frac{3}{2}\sigma}\right) +$$

$$+ \frac{1}{2} e^{-\frac{3}{4}\sigma} .$$

Proof: The characteristic function of S_n is shown to be

$$\varphi_{S_n}(t) = E\left(e^{iS_n t}\right) = \prod_{k=1}^{n} \left[1 - \frac{p_k}{q_k} (e^{it}-1)\right]^{-1}, \tag{2.1}$$

since the characteristic function of X_k is

$$\varphi_{X_k}(t) = \sum_{j=0}^{\infty} e^{ijt} \quad p_k^j q_k = \left[1 + \frac{p_k}{q_k} (1-e^{it})\right]^{-1}.$$

By Taylor's expansion of $\log \varphi_{\frac{S_n-\mu}{\sigma}}(t)$, where $\mu = E(S_n) = \sum_{k=1}^{n} \frac{p_k}{q_k}$,

$\sigma = D(S_n) = \left(\sum_{k=1}^{n} \frac{p_k}{q_k^2}\right)^{\frac{1}{2}}$, we have

$$\log \varphi_{\frac{S_n-\mu}{\sigma}}(t) = - \sum_{k=1}^{n} \log\left[1 - \frac{p_k}{q_k}(e^{\frac{it}{\sigma}}-1)\right] - i\mu \frac{t}{\sigma}$$

$$= \left(\sum_{k=1}^{n} \frac{p_k}{q_k}\right)\left(e^{\frac{it}{\sigma}}-1\right) + \left(\sum_{k=1}^{n} \frac{p_k^2}{q_k^2}\right)\left(e^{\frac{it}{\sigma}}-1\right)^2 + \dots$$

$$+ \left(\sum_{k=1}^{n} \frac{p_k^4}{q_k^4}\right)\left(e^{\frac{it}{\sigma}}-1\right)^4 + \sum_{k=1} \int_0^{\frac{p_k}{q_k}\left(e^{\frac{it}{\sigma}}-1\right)} \frac{z^4}{1-z}\, dz - i\mu\frac{t}{\sigma} . \tag{2.2}$$

In (2.2) substitute

$$e^{\frac{it}{\sigma}} = 1 + \frac{it}{\sigma} + \dots + \frac{1}{120}\left|\frac{t}{\sigma}\right|^5 \nabla, \dots, \left(e^{\frac{it}{\sigma}}-1\right)^4 = \frac{1}{4}\left(\frac{it}{\sigma}\right)^4 + \frac{137}{30}\left|\frac{t}{\sigma}\right|^5 \nabla,$$

where ∇'s are some functions of t such that $|\nabla| \leq 1$. Then we have

$$\log \varphi_{\frac{S_n-\mu}{\sigma}}(t) = - \frac{t^2}{2} + A\left(\frac{it}{\sigma}\right)^3 + B\left(\frac{it}{4}\right)^4 + R , \tag{2.3}$$

where $A = \sum_{k=1}^{n} \frac{1+p_k}{6q_k}\left(\frac{p_k}{q_k^2}\right)$, $B = \sum_{k=1}^{n} \frac{1+4p_k+p_k^2}{24q_k^2}\left(\frac{p_k}{q_k^2}\right)$ and

$$R = \sum_{k=1}^{n}\left(\frac{1}{120}\frac{p_k}{q_k} + \frac{17}{120}\frac{p_k^2}{q_k^2} + \frac{57}{60}\frac{p_k^3}{q_k^3} + \frac{137}{30}\frac{p_k^4}{q_k^4} + \frac{1}{5}\frac{p_k^5}{q_k^5}\right)\left|\frac{t}{\sigma}\right|^5 \nabla,$$

since

$$\left|\sum_{k=1}^{n}\int_0^{\frac{p_k}{q_k}\left(e^{\frac{it}{\sigma}}-1\right)} \frac{z^4}{1-z}\,dz\right| \leqq \sum_{k=1}^{n}\frac{1}{\min\limits_{k,t,z\varepsilon C}|1-z|}\int_0^{\frac{p_k}{q_k}\left(e^{\frac{it}{\sigma}}-1\right)} |z|^4\,|dz|$$

$$\leqq \frac{1}{5}\sum_{k=1}^{n}\frac{p_k^5}{q_k^5}\left|\frac{t}{\sigma}\right|^5,$$

where C denotes the path from 0 to $\frac{p_k}{q_k}\left(e^{\frac{it}{\sigma}}-1\right)$.

From (2.3) we have $\varphi_{\frac{S_n-\mu}{\sigma}}(t) = e^{-\frac{t^2}{2} + A\frac{it}{\sigma}^3 + B\frac{it}{}^4 + R}$. (2.4)

This relation and the Taylor expansion of $\frac{1}{\sin\frac{t}{2\sigma}}$ (see Lemma 3, p. 50, [6]) together imply that

$$\varphi_{\frac{S_n-\mu}{\sigma}}(t)\left(\frac{1}{\sin\frac{2t}{\sigma}}\right) = e^{-t^2/2}\left[\frac{2}{t} + 2Ai\left(\frac{it}{\sigma}\right)^2 + A^2 i\left(\frac{it}{\sigma}\right)^5 + 2Bi\left(\frac{it}{\sigma}\right)^3 + \frac{t}{12\sigma} + \Gamma\right], \quad (2.5)$$

where

$$\Gamma = \frac{A^3}{3}\,i\left(\frac{it}{\sigma}\right)^8 + \frac{A^4}{12}\left|\frac{t}{\sigma}\right|^{11} + 2ABi\left(\frac{it}{\sigma}\right)^6 + A^2B\left|\frac{t}{\sigma}\right|^9 + B^2\left|\frac{t}{\sigma}\right|^7\left(e^{B\left|\frac{t}{\sigma}\right|^4}\right) +$$

$$|R|\;\left|\frac{2\sigma}{t}\right|e^{|R|} + \left(|R|\right)\left(2B\left|\frac{t}{\sigma}\right|^3\right)e^{|R|+B\left|\frac{t}{\sigma}\right|^4} + \frac{7}{2880}\left|\frac{t}{\sigma}\right|^3 + \frac{A}{12}\left|\frac{t}{\sigma}\right|^4 +$$

$$\frac{7}{2880}A\left|\frac{t}{\sigma}\right|^6 + \left(\frac{A^2}{2}\right)\frac{1}{12}\left|\frac{t}{\sigma}\right|^7 + \frac{B}{12}\left|\frac{t}{\sigma}\right|^5 e^{B\left|\frac{t}{\sigma}\right|^4} + \left(|R|\right)\frac{1}{12}\left|\frac{t}{\sigma}\right|e^{|R|} +$$

$$\left(|R|\right)\left(\frac{B}{12}\left|\frac{t}{\sigma}\right|^5\right)e^{|R|+B\left|\frac{t}{\sigma}\right|^4}. \tag{2.6}$$

We shall use (2.5) as Taylor expansion for small t, and for large t we need the following evaluation of $\varphi_{\frac{S_n-\mu}{\sigma}}(t)$:

From (2.1) we have

$$\varphi_{\frac{S_n-\mu}{\sigma}}(t) = e^{-i\mu\frac{t}{\sigma}}\prod_{k=1}^{n}\left[1+\frac{p_k}{q_k}\left(1-e^{i\frac{t}{\sigma}}\right)\right]^{-1} = e^{-i\mu\frac{t}{\sigma}}\prod_{k=1}^{n}K_k,$$

$$K_k = \frac{1}{1+\frac{p_k}{q_k}\left(1-e^{i\frac{t}{\sigma}}\right)} = \frac{q_k}{1-p_k e^{\frac{it}{\sigma}}}. \tag{2.7}$$

Since

$$|K_k| = \frac{q_k}{\left|1-p_k e^{\frac{it}{\sigma}}\right|} = \frac{1}{\left[1+\frac{2p_k}{q_k^2}\left(1-\cos\frac{t}{\sigma}\right)\right]^{\frac{1}{2}}},$$

we have, assuming $\frac{p_k}{q_k^2} < \frac{3}{4}$ or $p_k < \frac{1}{3}$,

$$\log|K_k| = -\frac{1}{2}\log(1+x_k) \leqq -\frac{x_k}{2}+\frac{x_k^2}{4},\quad x_k = \frac{2p_k}{q_k^2}\left(1-\cos\frac{t}{\sigma}\right) \geqq 0.$$

Further, imposing strict restriction $|x_k| \leqq 1$, which is equivalent to $p_k \leqq 0.17$, we have

$$|K_k| \leqq e^{-x_k/4}, \tag{2.8}$$

from which we have

$$\left|\varphi_{\frac{S_n-\mu}{\sigma}}(t)\right| \leqq \prod_{k=1}^{n} e^{-\frac{x_k}{4}} \leqq e^{-\frac{1}{4}\left(\sum_{k=1}^{n}\frac{2p_k}{q_k^2}\right)\frac{2t^2}{\pi^2\sigma^2}} = e^{-\frac{t^2}{\pi^2}} \text{ or } e^{-\frac{t^2}{4}+\frac{t^4}{48\sigma^2}}, \tag{2.9}$$

by the relation $\left|1-\cos\frac{t}{\sigma}\right| = 2\sin^2\left(\frac{t}{2\sigma}\right) \geqq \frac{2t^2}{\pi^2\sigma^2}$ or $\frac{t^2}{2\sigma^2}-\frac{t^4}{24\sigma^4}$.

By Lévy's inversion formula, we have

$$\sum_{j=\ell+1}^{m} P(S_n=j) = \frac{1}{2\pi\sigma}\int_{-\pi\sigma}^{\pi\sigma} \varphi_{\frac{S_n-\mu}{\sigma}}(t)\, \frac{e^{-i\xi_1 t}-e^{-i\xi_2 t}}{2i\sin\frac{t}{2\sigma}}\, dt$$

$$= \frac{1}{2\pi\sigma}\left(\int_{-\beta\sqrt{\sigma}}^{\beta\sqrt{\sigma}} + \int_{\pi\sigma>|t|>\beta\sqrt{\sigma}}\right)\varphi_{\frac{S_n-\mu}{\sigma}}(t)\, \frac{e^{-i\xi t}-e^{i\xi_2 t}}{2i\sin\frac{t}{2\sigma}}\, dt = J_1 + J_2, \tag{2.10}$$

where $\xi_1 = \dfrac{\ell+\frac{1}{2}-\mu}{\sigma}$, $\xi_2 = \dfrac{m+\frac{1}{2}-\mu}{\sigma}$

For the calculus of J_1, we have from (2.5)

$$J_1 = \frac{1}{2\pi\sigma}\int_{-\beta\sqrt{\sigma}}^{\beta\sqrt{\sigma}} \left[\frac{\sigma}{it} + A\left(\frac{it}{\sigma}\right)^2\right] e^{-t^2/2}\left(e^{-i\xi_1 t}-e^{-i\xi_2 t}\right) dt$$

$$+ \frac{1}{2\pi\sigma}\int_{-\beta\sqrt{\sigma}}^{\beta\sqrt{\sigma}} \left[A^2\left(\frac{it}{\sigma}\right)^5 + 2B\left(\frac{it}{\sigma}\right)^3 - \frac{1}{12}\left(\frac{it}{\sigma}\right)\right] e^{-t^2/2}\left(e^{-i\xi_1 t}-e^{-i\xi_2 t}\right) dt$$

$$+ \frac{1}{2\pi\sigma}\int_{-\beta\sqrt{\sigma}}^{\beta\sqrt{\sigma}} \Gamma e^{-t^2/2}\left(e^{-i\xi_1 t}-e^{-i\xi_2 t}\right) dt$$

$$= \frac{1}{2\pi\sigma}\int_{-\infty}^{\infty} \left[\frac{\sigma}{it} + A\left(\frac{it}{\sigma}\right)^2\right] e^{-t^2/2}\left(e^{-i\xi_1 t}-e^{i\xi_2 t}\right) dt$$

$$- \frac{1}{2\pi\sigma}\int_{|t|>\beta\sqrt{\sigma}} \left[\frac{\sigma}{it} + A\left(\frac{it}{\sigma}\right)^2\right] e^{-t^2/2}\left(e^{i\xi_1 t}-e^{i\xi_2 t}\right) dt + \Gamma_2 + \Gamma_3$$

$$= \int_{\xi_1}^{\xi_2} \frac{1}{\sqrt{2\pi}} e^{-\frac{u^2}{2}} du + \frac{A}{\sqrt{2\pi\sigma}}\left[\left(1-\xi_2^2\right)e^{-\frac{\xi_2^2}{2}} - \left(1-\xi_2^2\right)e^{-\frac{\xi_1^2}{2}}\right]$$

$$+ \Gamma_1 + \Gamma_2 + \Gamma_3. \tag{2.11}$$

For J_2 we have, from (2.8), putting $\beta = \sqrt{3\sigma}$:

$$|J_2| \leqq \frac{2}{2\pi\sigma}\int_{\sqrt{3\sigma}\,\frac{\pi}{2}}^{\infty}\left(e^{-\frac{t^2}{\pi^2}}\Big/\frac{t}{\pi\sigma}\right)dt + \frac{2}{2\pi\sigma}\int_{\sqrt{3\sigma}}^{\sqrt{3\sigma}\,\frac{\pi}{2}}\left(e^{-\frac{t^2}{4}+\frac{t^4}{48}}\Big/\frac{t}{\pi\sigma}\right)dt$$

$$\leqq \left(\frac{2}{3\sigma} + e^{\frac{9}{48}}\log\frac{\pi}{2}\right)e^{-\frac{3}{4}\sigma} \leqq \left(\frac{2}{3\sigma}+\frac{1}{2}\right)e^{-\frac{3}{4}\sigma}. \qquad (2.12)$$

Thus, it suffices to bound Γ_1, Γ_2, and Γ_3. First, we have

$$|\Gamma_1| \leqq \frac{2}{\pi}\int_{\sqrt{3\sigma}}^{\infty}\frac{e^{-t^2/2}}{t}\,dt + \frac{1}{2\pi\sigma^3}\int_{\sqrt{3\sigma}}^{\infty}t^2e^{-t^2/2}\,dt \leqq \frac{2}{3\sigma}\,e^{-\frac{3}{2}\sigma}, \quad (2.13)$$

where A is bounded by $\frac{1}{4}$ for $p_k < 0.17$. For Γ_2 we have

$$|\Gamma_2| \leqq \frac{1}{2\pi\sigma}\int_{-\beta\sqrt{\sigma}}^{\beta\sqrt{\sigma}}\left|\left[A^2\left(\frac{it}{\sigma}\right)^5 + 2B\left(\frac{it}{\sigma}\right)^3 - \frac{1}{12}\left(\frac{it}{\sigma}\right)\right]e^{-t^2/2}\,\frac{\left(e^{-i\xi_1t}-e^{-i\xi_2t}\right)}{2i}\right|dt$$

$$\leqq \frac{1}{\pi\sigma}\int_0^{\infty}\left|\frac{1}{\sigma}\left(\frac{A^2}{\sigma^4}t^5 - \frac{2B'}{\sigma^2}t^3\right)\right|e^{-t^2/2}\,dt +$$

$$\frac{1}{\pi\sigma}\int_0^{\infty}\left[\frac{1}{\sigma}\left(\frac{2B''}{\sigma^2}\,t^3 + \frac{1}{12\sigma}\,t\right)\right]e^{-t^2/2}\,dt = \Gamma_{21} + \Gamma_{22}, \qquad (2.14)$$

where $B' = \sum_{k=1}^{n}\frac{1+2p_k+p_k^2}{24q_k^2}\,\frac{p_k}{q_k^2}$, $B'' = \sum_{k=1}^{n}\frac{2p_k}{24q_k^2}\,\frac{p_k}{q_k^2}$. Putting

$a = \frac{1+p_j}{2q_j}\Big/\frac{1+p_k}{6q_j}$, which is bounded by 4.23 for $p_k \leqq 0.17$,

$(k = 1,2,\ldots,n)$, we have for $\sigma \leqq 5$:

$$\Gamma_{21} \leqq \frac{1}{\sigma 6\pi}\int_0^{\infty}\left|\sum_k\frac{p_k}{q_k^2}\sum_j\frac{1+p_j}{6q_j}\left(\frac{p_j}{q_j^2}\right)\left(\frac{1+p_k}{6q_k}\,t^5 - \frac{1+p_j}{2q_j}\,t^3\right)\right|e^{-t^2/2}\,dt$$

$$\leqq \frac{1}{\sigma^2}\,\frac{(0.235)^2}{\pi}\max_{0\leqq a\leqq 4.23}\frac{a}{3}\int_0^{\infty}|t^5-at^3|e^{-t^2/2}\,dt$$

$$= \frac{0.06}{\sigma^2}$$

and

$$\Gamma_{22} = \frac{1}{\pi\sigma^3}\left(\frac{2B''}{\sigma}\int_0^\infty t^3 e^{-t^2/2}\,dt + \frac{1}{12}\int_0^\infty t e^{-t^2/2}\,dt\right) \leqq \frac{0.11}{\sigma^2}.$$

These imply that

$$|\Gamma_2| \leqq \frac{0.17}{\sigma^2}. \tag{2.15}$$

We can evaluate the boundary of Γ_3 by inequalities $A \leqq \frac{\sigma^2}{4}$, $B \leqq \frac{\sigma^2}{10}$, and $|\Gamma| \leqq 0.025\left|\frac{t}{\sigma}\right|^5$ when $p_k \leqq 0.17$ $(k = 1,2,\ldots,n)$; and we have

$$|\Gamma_3| \leqq \frac{1}{\pi\sigma}\int_0^\infty \left(\frac{A^3}{3}\frac{t^8}{\sigma^8} + \frac{A^4}{12}\frac{t^{11}}{\sigma^{11}} + 2AB\frac{t^6}{\sigma^6} + A^2B\frac{t^9}{\sigma^9} + B^2\frac{t^7}{\sigma^7}\left(e^{B\frac{\beta^4\sigma^2}{\sigma^4}}\right) +\right.$$

$$2(0.025)\frac{t^4}{\sigma^4}e^{0.025\frac{\beta^5\sigma^{5/2}}{\sigma^5}} + 0.025\,(2B)\,\frac{t^8}{\sigma^8}e^{B\frac{\beta^4\sigma^2}{\sigma^4}+0.025\frac{\beta^5\sigma^{5/2}}{\sigma^5}} +$$

$$\frac{7}{2880}\frac{t^3}{\sigma^3} + \frac{A}{12}\frac{t^4}{\sigma^4} + \frac{7}{2880}A\frac{t^6}{\sigma^6} + \frac{A^2}{2}\;\frac{1}{12}\frac{t^7}{\sigma^7} + \frac{B}{12}\frac{t^5}{\sigma^5}e^{B\frac{\beta^4\sigma^2}{4}} +$$

$$(0.025)\,\frac{1}{12}\frac{t^6}{\sigma^6}e^{0.025\frac{\beta^5\sigma^{5/2}}{\sigma^5}} +$$

$$\left.(0.025)\,\frac{B}{12}\frac{t^{10}}{\sigma^{10}}e^{B\frac{\beta^4\sigma^2}{\sigma^4}+0.025\frac{\beta^5\sigma^{5/2}}{\sigma^5}}\right)e^{-t^2/2}\,dt$$

$$\leqq \frac{0.51}{\sigma^3} + \frac{0.99}{\sigma^4} + \frac{0.13}{\sigma^5} + \frac{1.09}{\sigma^7} + \frac{0.31}{\sigma^9}\,. \tag{2.16}$$

3. POISSON APPROXIMATION

Since the characteristic function of S_n is expressed as

$$E\left(e^{iS_n t}\right) = \varphi_{S_n}(t) = \prod_{k=1}^{n}\left[1 + \frac{p_k}{q_k}(1-e^{it})\right]^{-1}, \tag{3.1}$$

we have by Taylor's expansion with remainder term

$$\log\varphi_{S_n}(t) = -\sum_{k=1}^{n}\log\left[1 - \frac{p_k}{q_k}(e^{it}-1)\right]$$

$$= \sum_{k=1}^{n}\frac{p_k}{q_k}(e^{it}-1) + \frac{1}{2}\sum_{k=1}^{n}\frac{p_k}{q_k^2}(e^{it}-1)^2 +$$

$$\frac{1}{3\min\limits_{1\leq k\leq n,\, t}\left|1-\frac{p_k}{q_k}(e^{it}-1)\right|}\sum_{k=1}^{n}\frac{p_k^3}{q_k^3}|e^{it}-1|^3\,\nabla,$$

where $|\nabla|\leq|$. (3.2)

Hence, we have by (3.2)

$$\varphi_{S_n}(t) = \exp\left[\lambda_1(e^{it}-1) + \frac{\lambda_2}{2}(e^{it}-1)^2 + \frac{\nabla}{\min\limits_{1\leq k\leq n,\, t}\left|1-\frac{p_k}{q_k}(e^{it}-1)\right|}\cdot\left(\frac{\lambda_3}{3}\right)|e^{it}-1|^3\right], \quad (3.3)$$

where $\lambda_i = \sum_{k=1}^{n}(p_k/qk)^i$, $i = 1,2,3$, from which it follows by Lévy's inversion formula

$$\sum_{j=\ell+1}^{m}P(S_n=j) = \frac{1}{2\pi}\int_{-\pi}^{\pi}\varphi_{S_n}(t)\sum_{j=\ell+1}^{m}e^{ijt}\,dt$$

$$= \frac{1}{2\pi}\int_{-\pi}^{\pi}\varphi_{S_n}(t)e^{-i(\ell+m+1)t/2}\,\frac{\sin(m-\ell)t/2}{\sin t/2}\,dt$$

$$= \frac{1}{2\pi}\int_{-\pi}^{\pi}e^{\lambda_1(e^{it}-1)}e^{-i(\ell+m+1)t/2}\left[\frac{\sin(m-\ell)t/2}{\sin t/2}\right]dt +$$

$$\frac{\lambda_2}{2}\,\frac{1}{2\pi}\int_{-\pi}^{\pi}\left\{\left[e^{\lambda_1(e^{it}-1)+2it} - e^{\lambda_1(e^{it}-1)+it}\right] - \left[e^{\lambda_1(e^{it}-1)+it} - e^{\lambda_1(e^{it}-1)}\right]\right\}\ \text{mult. by}$$

$$\left\{e^{-i(\ell+m+1)t/2}\left[\frac{\sin(m-\ell)t/2}{\sin t/2}\right]dt\right\} + \tilde{\P}, \quad (3.4)$$

where

$$|\tilde{\P}| \leqq \frac{1}{2\pi}\int_{-\pi}^{\pi} \frac{\lambda_2^2}{8}|e^{it}-1|^4 |e^{\lambda_1(e^{it}-1)}| \, |e^{\frac{\lambda_2}{2}|e^{it}-1|^2}| \frac{dt}{|\sin t/2|} +$$

$$\frac{1}{2\pi}\int_{\pi}^{\pi} \frac{\lambda_3}{3} \frac{|e^{it}-1|^3}{\min\limits_k\left|1-2\frac{p_k}{q_k}\right|} \exp\left(\frac{\lambda_3}{3}\frac{|e^{it}-1|^3}{\min\limits_k\left|1-2\frac{p_k}{q_k}\right|}\right)\left|e^{\lambda_1(e^{it}-1)}\right|\left|e^{\frac{\lambda_2}{2}(e^{it}-1)}\right| \cdot$$

$$\frac{dt}{|\sin t/2|}$$

$$= R_1 + R_2.$$

We can evaluate R_1 and R_2 as follows:

$$R_1 \leqq \frac{\lambda_2^2}{8}\frac{1}{\pi}\int_0^{\pi} t^4 \, e^{-2\lambda_1 \sin^2 \frac{t}{2} + \frac{\lambda_2}{2}t^2} \frac{dt}{t/\pi}$$

$$\leqq \frac{\lambda_2^2}{8}\int_0^{\infty} t^3 \, e^{-\frac{(\sqrt{\lambda_1}t)^2}{2}\left(\frac{4}{\pi^2} - \frac{\lambda_2}{\lambda_1}\right)} dt$$

$$\leqq \frac{1}{4}\left(\frac{\lambda_2}{\lambda_1}\right)^2 \cdot \frac{1}{\left(\frac{4}{\pi^2} - \frac{\lambda_2}{\lambda_1}\right)^2} \leqq \frac{\alpha^2}{4} \frac{1}{\left(\frac{4}{\pi^2} - \alpha\right)^2}, \tag{3.5}$$

where

$$\alpha = \max_k \frac{p_k}{q_k},$$

and

$$R_2 \leqq \frac{\lambda_3}{3}\frac{1}{\pi}\int_0^{\pi} \frac{t^3}{\min\limits_k\left|1-2\frac{p_k}{q_k}\right|} \exp\left[\frac{2t^2}{3\min\limits_k\left|1-2\frac{p_k}{q_k}\right|} - \frac{(\sqrt{\lambda}t)^2}{2}\left(\frac{4}{\pi^2} - \frac{\lambda_2}{\lambda_1}\right)\right] \frac{dt}{t/\pi}$$

$$\leqq \frac{\lambda_3}{3(1-2\alpha)}\int_0^{\infty} t^2 \, e^{-\frac{1}{2}\left[\lambda_1\left(\frac{4}{\pi^2} - \frac{\lambda_2}{\lambda_1}\right) - \frac{4}{3}\frac{\lambda_3}{1-2\alpha}\right]t^2} dt.$$

Since $\int_0^\infty t^2 e^{-\frac{1}{t}at^2} dt = (\pi/2a^3)^{\frac{1}{2}}$, it follows that

$$R_2 \leqq \frac{\lambda_3}{3(1-2\alpha)} \frac{\sqrt{\pi}}{\sqrt{2}\left[\lambda_1\left(\frac{4}{\pi^2} - \frac{\lambda_2}{\lambda_1}\right) - \frac{4}{3}\frac{\lambda_3}{1-2\alpha}\right]^{3/2}} \cdot$$

$$\leqq \frac{1}{3}\left(\frac{\pi}{2}\right)^{\frac{1}{2}} \frac{\alpha^2}{(1-2\alpha)\lambda_1^{1/2}} \cdot \frac{1}{\left[\frac{4}{\pi^2} - \frac{3-2\alpha}{3(1-2\alpha)}\alpha\right]^{3/2}} \cdot \tag{3.6}$$

If one wishes to obtain a boundary for R_2 free of λ_1, we can estimate R_2 as

$$R_2 \leqq \frac{\lambda_3}{3}\frac{1}{\pi}\int_0^\pi \frac{2t^2}{\min_k\left|1-2\frac{p_k}{q_k}\right|} e^{-\frac{1}{2}\left[\left(\frac{4}{\pi^2} - \frac{\lambda_2}{\lambda_1}\right) - \frac{4}{3}\frac{3}{1-2\alpha}\right](\sqrt{\lambda}_1 t)^2} \frac{dt}{t/\pi}$$

$$\leqq \frac{2\alpha^2}{3(1-2\alpha)\left[\frac{4}{\pi^2} - \frac{3-2\alpha}{3(1-2\alpha)}\alpha\right]} \cdot \tag{3.7}$$

Since the first and the second terms of the last expression of (3.4) are, respectively, $\sum_{j=\ell+1}^{m} \frac{\lambda_1^j e^{-\lambda_1}}{j!}$ and $\frac{\lambda_2}{2}\left[\Delta^2 P(m) - \Delta^2 P(\ell)\right]$, where

$$\Delta^2 P(j') = \left[P(j')-P(j'-1)\right] - \left[P(j'-1)-P(j'-2)\right], P(j') = \sum_{j=0}^{j'} \frac{\lambda_1^j e^{-\lambda_1}}{j!}, \tag{3.8}$$

we can summarize the above discussions in the following theorem:

<u>Theorem 3.1</u>: Let S_n be defined as the generalized negative binomial random variable, and $\alpha = \max_{1\leqq k\leqq n} \frac{p_k}{q_k}$ is small and is restricted as $\alpha \leqq \frac{2}{5}$ or $p_k \leqq \frac{2}{7}$, then we have

$$\sum_{j=\ell+1}^{m} P(S_n=j) = \sum_{j=\ell+1}^{m} \frac{\lambda_1^j e^{-\lambda}}{j!} + \frac{\lambda_2}{2}\left[\Delta^2 P(m) - \Delta^2 P(\ell)\right] + \tilde{\P} \tag{3.9}$$

where

$$|\tilde{\P}| \leqq \alpha^2 \left\{ \frac{1}{4}\left(\frac{4}{\pi^2} - \alpha\right)^{-1} + \frac{\sqrt{\pi}}{3\sqrt{2}} \lambda_1^{-\frac{1}{2}} (1-2\alpha)^{-1} \left[\frac{4}{\pi^2} - \frac{3-2\alpha}{3(1-2\alpha)}\alpha\right]^{-\frac{3}{2}} \right\}$$

or

$$\leqq \alpha^2 \left\{ \frac{1}{4}\left(\frac{4}{\pi^2} - \alpha\right)^{-1} + \frac{2}{3}(1-2\alpha)^{-1} \left[\frac{4}{\pi^2} - \frac{3-2\alpha}{3(1-2\alpha)}\alpha\right]^{-1} \right\} \tag{3.10}$$

4. NEGATIVE BINOMIAL APPROXIMATION

In Section 3 Poisson approximation formula is derived with correction term of second order difference. Also it is necessary to obtain the formula which approximates the generalized negative binomial distribution with error term of order $O(\alpha^2)$ and without correction term. In this light, we see in (3.2) that the second term of the last expression should be identified as the following expansion of negative binomial characteristic function $\varphi_{NB}(t)$:

$$\log\varphi_{NB}(t) = -\frac{\lambda}{\rho}\log[1-\rho(e^{it}-1)]$$

$$= \lambda(e^{it}-1) + \frac{\lambda\rho}{2}(e^{it}-1)^2 + \frac{\lambda\rho^2}{3}(e^{it}-1)^3 + \ldots, \tag{4.1}$$

where $\lambda = \lambda_1 = \sum_{k=1}^{n} \frac{p_k}{q_k}$, $\lambda\rho = \lambda_2 = \sum_{k=1}^{n} \frac{p_k^2}{q_k^2}$. From (3.3) we have

further, $\varphi_{S_n}(t) = \exp\left[\lambda_1(e^{it}-1) + \frac{\lambda_2}{2}(e^{it}-1)^2 + \frac{\lambda_3}{3}(e^{it}-1)^3 + \right.$

$$\left. \frac{\lambda_4}{4\min\limits_{1\leqq k\leqq n,\, t}\left|1-\frac{p_k}{q_k}(e^{it}-1)\right|}|e^{it}-1|^4\nabla\right]$$

$$= \exp\left[\lambda(e^{it}-1) + \frac{\lambda\rho}{2}(e^{it}-1)^2 + \frac{\lambda\rho^2}{3}(e^{it}-1)^3 + \frac{\lambda\rho^3}{4}(e^{it}-1)^4 + \ldots\right.$$

$$+ \frac{\lambda}{\rho}\frac{\rho^4}{4\min\limits_{t}|1-\rho(e^{it}-1)|}|e^{it}-1|^4\,\nabla + \frac{1}{3}(\lambda_3-\lambda\rho^2)(e^{it}-1)^3 +$$

$$\left.\frac{\lambda_4}{4\min\left|1-\frac{p_k}{q_k}(e^{it}-1)\right|}|e^{it}-1|^4\,\nabla\right]$$

$$= [1-\rho(e^{it}-1)]^{-\frac{\lambda}{\rho}} \; e^{\frac{1}{3}(\lambda_3-\lambda\rho^2)(e^{it}-1)^3+R} \; , \qquad (4.2)$$

where R is the remainder term. Hence, we have by Taylor's expansion of (4.2)

$$\varphi_{S_n}(t) = [1-\rho(e^{it}-1)]^{-\frac{\lambda}{\rho}} + [1-\rho(e^{it}-1)]^{-\frac{\lambda}{}} \cdot \frac{\lambda_3-\lambda\rho^2}{3}(e^{it}-1)^3 +$$

$$|\varphi_{NB}(t)| \cdot \frac{(\lambda_3-\lambda\rho^2)^2}{18} \, |e^{it}-1|^6 \, e^{\frac{|\lambda_3-\lambda\rho^2|}{3} \cdot |e^{it}-1|^3} \, \nabla +$$

$$|\varphi_{NB}(t)| \cdot \left| e^{\frac{1}{3}(\lambda_3-\lambda\rho^2)(e^{it}-1)^3} \right| \cdot |R| e^{|R|} \, \nabla \, . \qquad (4.3)$$

From Lévy's inversion formula,

$$\sum_{j=\ell+1}^{m} P(S_n=j) = \frac{1}{2\pi}\int_{-\pi}^{\pi} \varphi_{S_n}(t) e^{-i(\ell+m+1)t/2}\left[\frac{\sin(m-\ell)t/2}{\sin t/2}\right] dt$$

$$= \frac{1}{2\pi}\int_{-\pi}^{\pi} \varphi_{NB}(t) e^{-i(\ell+m+1)t/2}\left[\frac{\sin(m-\ell)t/2}{\sin t/2}\right] dt \; +$$

$$\frac{\lambda_3-\lambda\rho^2}{3} \cdot \frac{1}{2\pi}\int_{-\pi}^{\pi} (e^{it}-1)^3 \varphi_{NB}(t) e^{-i(\ell+m+1)t/2} \; \cdot$$

$$\frac{\sin(m-\ell)t/2}{\sin t/2}\, dt + \P^*, \qquad (4.4)$$

where ¶* is Lévy's inversion of the remainder term in (4.3).

Now we take up estimation of ¶*; from (4.3) we have

$$|\P^*| \leqq \frac{(\lambda_3-\lambda\rho^2)^2}{18} \, \frac{1}{2\pi}\int_{-\pi}^{\pi} 4t^4 e^{\frac{2|\lambda_3-\lambda\rho^2|}{3} \cdot t^2 - \left(\frac{4}{\pi^2} - \frac{\rho}{1-2\rho}\right)\frac{\lambda t^2}{2}} \frac{dt}{|t|/\pi} \; +$$

$$\frac{1}{2\pi}\int_{-\pi}^{\pi} 4t^2 \cdot \frac{1}{4} \left| \frac{\lambda\rho^3}{1-2\rho} + \frac{\lambda_4}{\min\limits_k \left|1-2\,\frac{p_k}{q_k}\right|} \right| \text{ mult. by}$$

$$\exp\left[\frac{2|\lambda_3-\lambda\rho^2|}{3}t^2 + \left(\frac{\lambda\rho^3}{1-2\rho} + \frac{\lambda_4}{\min\left|1-2\,\frac{p_k}{q_k}\right|}\right)t^2 - \left(\frac{4}{\pi^2} - \frac{\rho}{1-2\rho}\right)\frac{\lambda t^2}{2}\right]\frac{dt}{|t|/\pi} = R_1^* + R_2^*. \tag{4.5}$$

Since

$$|\log\varphi_{NB}(t) - \lambda(e^{it}-1)| \leq \frac{\lambda\rho t^2}{2\min_t|1-\rho(e^{it}-1)|} \leq \frac{\lambda\rho t^2}{2|1-2\rho|},$$

which implies

$$|\varphi_{NB}(t)| \leq \left|e^{-\lambda(e^{it}-1)}\right| \cdot e^{\frac{\lambda\rho t^2}{2(1-2\rho)}} \leq e^{-\left(\frac{4}{\pi^2} - \frac{\rho}{1-2\rho}\right)\frac{\lambda t^2}{2}}.$$

Using arguments similar to those of Section 3, we have

$$R_1^* \leq \frac{2(\lambda_3-\lambda\rho^2)^2}{9}\int_0^\infty t^3\, e^{-\frac{(\sqrt{\lambda}t)^2}{2A}}\,dt$$

$$= \frac{2}{9}\left(\frac{\lambda_3-\lambda\rho^2}{\lambda}\right)^2\int_0^\infty (\sqrt{\lambda}t)^3\, e^{-\frac{(\sqrt{\lambda}t)^2}{2A}}\sqrt{\lambda}\,dt$$

$$= \frac{8}{9}\left(\frac{\lambda_3-\lambda\rho^2}{\lambda}\right)^2 A^2, \tag{4.6}$$

and

$$R_2^* \leq \frac{a'}{\lambda}\int_0^\infty (\sqrt{\lambda}t)e^{-\frac{(\sqrt{\lambda}t)^2}{2B}}\sqrt{\lambda}\,dt = \frac{a'}{\lambda}B, \tag{4.7}$$

where

$$A = \left[\left(\frac{4}{\pi^2} - \frac{\rho}{1-2\rho}\right) - \frac{4}{3\lambda}|\lambda_3-\lambda\rho^2|\right]^{-1}, \quad a' = \frac{\lambda\rho^3}{1-2\rho} + \frac{\lambda_4}{\min_k\left|1-2\,\frac{p_k}{q_k}\right|}$$

$$B = \left[\left(\frac{4}{\pi^2} - \frac{\rho}{1-2\rho}\right) - \frac{2}{\lambda}\left(\frac{\lambda\rho^3}{1-2\rho} + \frac{\lambda_4}{\min_k\left|1-2\,\frac{p_k}{q_k}\right|}\right) - \frac{4}{3\lambda}|\lambda_3-\lambda\rho^2|\right]^{-1}.$$

Assuming that $\alpha = \max_k \frac{p_k}{q_k}$ is small, we have $\frac{1}{\lambda}|\lambda_3 - \lambda\rho^2| \leqq 2\alpha^2$ and $\frac{\alpha}{\lambda} \leqq \frac{2\alpha^3}{1-2\alpha}$. Hence, it follows that

$$R_1^* \leqq \frac{32}{9}\alpha^4\left(\frac{4}{\pi^2} - \frac{\alpha}{1-2\alpha} - \frac{8}{3}\alpha^2\right)^{-2}$$

and (4.8)

$$R_2^* \leqq \frac{2\alpha^3}{1-2\alpha}\left(\frac{4}{\pi^2} - \frac{\alpha}{1-2\alpha} - \frac{8}{3}\alpha^2 - \frac{4\alpha^3}{1-2\alpha}\right)^{-1} .$$

We can also state that the first and second terms of the last expression of (4.4) are, resp., $\sum_{j=\ell+1}^{m}\binom{\frac{\lambda}{\rho}+j+1}{j}\left(\frac{1}{1+\rho}\right)^{\frac{\lambda}{\rho}}\left(\frac{\rho}{1+\rho}\right)^{j}$ and $\frac{\lambda_3-\lambda\rho^2}{3}\{\Delta^3\tilde{P}(m)-\Delta^3\tilde{P}(\ell)\}$, where $P(\tilde{k}) = \sum_{j=0}^{k}\binom{\frac{\lambda}{\rho}+j-1}{j}\left(\frac{1}{1+\rho}\right)^{\frac{\lambda}{\rho}}\left(\frac{\rho}{1+\rho}\right)^{j}$.

Thus we have the following theorem:

<u>Theorem 4.2</u>: Let S_n be as defined in Theorem 4.1, and $\alpha = \max_k \frac{p_k}{q_k}$ is small (e.g., $\alpha \leq 0.17$); then we have

$$\sum_{j=\ell+1}^{m} P(S_n=j) = \sum_{j=\ell+1}^{m}\binom{\frac{\lambda}{\rho}+j-1}{j}\left(\frac{1}{1+\rho}\right)^{\frac{\lambda}{\rho}}\left(\frac{\rho}{1+\rho}\right)^{j} + \frac{\lambda_3-\lambda\rho^2}{3}[\Delta^3\tilde{P}(m)-\Delta^3\tilde{P}(\ell)] + \P^*, \tag{4.9}$$

where

$$|\P^*| \leqq \frac{2\alpha^3}{1-2\alpha}\left(\frac{4}{\pi^2} - \frac{\alpha}{1-2\alpha} - \frac{8}{3}\alpha^2 - \frac{4\alpha^3}{1-2\alpha}\right)^{-1} + \frac{32}{9}\alpha^4\left(\frac{4}{\pi^2} - \frac{\alpha}{1-2\alpha} - \frac{8}{3}\alpha^2\right)^{-2}. \tag{4.10}$$

5. <u>APPROXIMATION BY INFINITELY DIVISIBLE LAW</u>

In this section we shall prove the following theorem:

<u>Theorem 5.1</u>: Let X_k be a discrete random variable such that $P(X_k \neq 0) = q_k$ is small, and let $\exp \sum_{k=1}^{n}\left[E\left(e^{iX_k t}\right)-1\right]$ be the characteristic function of some discrete infinitely divisible law;

then we have

$$\left| \sum_{j=\ell+1}^{m} P(S_n = \sum_{k=1}^{n} X_k = j) - \sum_{j=\ell+1}^{m} P(S_n'=j) \right|$$

$$\leq \frac{\pi^2}{8}\tilde{\alpha} + \left[\left(\frac{4}{\pi^2} - \tilde{\alpha} \right)^{-2} + \frac{1}{3}(1-2\tilde{\alpha})^{-1} \left(\frac{2}{\pi^2} - \frac{\tilde{\alpha}}{2} - \frac{2\tilde{\alpha}^2}{3(1-2\tilde{\alpha})} \right)^{-1} \right] \tilde{\alpha}^2,$$

where $\tilde{\alpha} = \max_k \alpha_k$ and S_n' is the random variable having

$\exp \sum_{k=1}^{n} \left[E\left(e^{iX_k t}\right) - 1 \right]$ for its characteristic function. In particular, if X_k is defined as $P(X_k=j) = q_k p_k^j$, $j = 0,1,2,\ldots$, then we have

$$\exp \sum_{k=1}^{n} \left[E\left(e^{iX_k t}\right) - 1 \right] = \exp \sum_{k=1}^{n} \frac{\frac{p_k}{q_k}(e^{it}-1)}{1 - \frac{p_k}{q_k}(e^{it}-1)}$$

$$= \exp\left[\sum_{k=1}^{n} \frac{p_k}{q_k}(e^{it}-1) + \sum_{k=1}^{n} \frac{p_k^2}{q_k^2}(e^{it}-1)^2 + \sum_{k=1}^{n} \frac{p_k^3}{q_k^3}(e^{it}-1)^3 + \ldots \right]. \tag{5.1}$$

Ignoring the correction term, we can state that the negative binomial approximation to generalized negative binomial distribution has error term of order $\tilde{\alpha}^2$, while Poisson approximation and approximation by infinitely divisible law have error term of order $\tilde{\alpha}$ (although in the last case approximation by infinitely divisible law does not make sense for geometrical distributed random variable X_k because the distribution of X_k is already infinitely divisible). But we have the following interesting properties of the generalized negative binomial distribution:

1) in Theorem 3.1, the sign of correction term is positive while in generalized positive binomial distribution (see Theorem 1, [7]) it is negative; and
2) since close examination of (5.1) shows that distribution

   of S_n' is approximated by $\sum_{j=\ell+1}^{m} \frac{\lambda_1^j e^{-\lambda_1}}{j!} + \lambda_2[\Delta^2 P(m) - \Delta^2 P(\ell)]$,

   ignoring error terms of order $\tilde{\alpha}^2$, the following relations are deduced:

Generalized Negative Binomial Distribution $\sum_{j=0}^{m} P(S_n = j)$

$$= \sum_{j=0}^{m} \binom{\frac{\lambda}{\rho}+j-1}{j} \left(\frac{1}{1+\rho}\right)^{\frac{\lambda}{\rho}} \left(\frac{\rho}{1+\rho}\right)^{j} \qquad \text{(Theorem 4.2)}$$

$$= \sum_{j=0}^{m} \frac{\lambda_1^{j} e^{-\lambda_1}}{j!} + \frac{\lambda_2}{2}\, \Delta^2 P(m) \qquad \text{(Theorem 4.1)}$$

$$= \sum_{j=0}^{m} P(S_n' = j) - \frac{\lambda_2}{2}\, \Delta^2 P(m). \qquad \text{(Theorem 5.1 and eq. (5.1))}$$

(Proof of Theorem 5.1)

Since characteristic function $\varphi_{S_n}(t)$ of S_n is

$$\varphi_{S_n}(t) = \prod_{k=1}^{n} E\left(e^{itX_k}\right)$$

$$= \prod_{k=1}^{n} \left[(1-\alpha_k) + \int_{x\neq 0} e^{itx} dF_k(x)\right], \tag{5.2}$$

we have

$$\log \varphi_{S_n}(t) = \sum_{k=1}^{n} \log\left[1 + \int_{-\infty}^{\infty} W\right], \text{ where } W = (e^{itx}-1)\,dF_k(x),$$

$$= \sum_{k=1}^{n} \int_{-\infty}^{\infty} W - \frac{1}{2} \sum_{k=1}^{n} \left[\int_{-\infty}^{\infty} W\right]^2 +$$

$$\sum_{k=1}^{n} \int_{0}^{\int_{-\infty}^{\infty} W} \frac{z^2 dz}{1-z} .$$

From this we have

$$\varphi_{S_n}(t) = e^{\sum_{k=1}^{n}\left[E\left(e^{itX_k}\right)-1\right]} \cdot \exp\left\{-\frac{1}{2} \sum_{k=1}^{n} \left[\int_{-\infty}^{\infty} W\right]^2 + \sum_{k=1}^{n} \int_{0}^{\int_{-\infty}^{\infty} W} \frac{z^2 dz}{1-z}\right\}$$

$$= e^{\sum_{k=1}^{n}\left[E\left(e^{itX_k}\right)-1\right]}\left[1 - \frac{1}{2}\sum_{k=1}^{n}\left(\int_{-\infty}^{\infty} W\right)^2 + \frac{\nabla}{8}\left|\sum_{k=1}^{n}\left(\int_{-\infty}^{\infty} W\right)^2\right|^2\right. \text{ mult. by}$$

$$e^{\frac{1}{2}\left|\sum_{k=1}\left(\int_{\infty}^{\infty} W\right)^2\right|}\Bigg] \cdot \left[1 + \left|\sum_{k=1}\int_0^{\int_{-\infty}^{\infty} W} \frac{z^2dz}{1-z}\right|\right. \text{ mult. by}$$

$$e^{\left|\sum_{k=1}^{n}\int_0^{\int_{-\infty}^{\infty} W} \frac{z^2dz}{1-z}\right|}\Bigg] = e^{\sum_{k=1}^{n}\left[E\left(e^{itX_k}\right)1\right]} - \frac{1}{2}\sum_{k=1}^{n}\left(\int_{-\infty}^{\infty} W\right)^2 \text{ mult. by}$$

$$e^{\sum_{k=1}^{n}\left[E\left(e^{itX_k}\right)-1\right]} + \tilde{R}_2 + \tilde{R}_3, \tag{5.3}$$

where

$$|\tilde{R}_2| \leq \frac{1}{8}\left|\sum_{k=1}^{n}\left(\int_{-\infty}^{\infty} W\right)^2\right| e^{\frac{1}{2}\left|\sum_{k=1}^{n}\left(\int_{-\infty}^{\infty} W\right)^2\right|} \cdot \left|e^{\sum_{k=1}^{n}\left[E\left(e^{itX_k}\right)-1\right]}\right|,$$

and

$$|\tilde{R}_3| \leq \left|\sum_{k=1}^{n}\int_0^{\int_{-\infty}^{\infty} W} \frac{z^2}{1-z}\,dz\right| \cdot e^{\left|\sum_{k=1}^{n}\int^{\int_{-\infty}^{\infty} W} \frac{z^2}{1-z}\,dz\right|} \cdot$$

$$\left|e^{\sum_{k=1}^{n}\left[E\left(e^{itX_k}\right)-1\right]}\right| \cdot \left|e^{-\frac{1}{2}\sum_{k=1}^{n}\left(\int_{-\infty}^{\infty} W\right)^2}\right|. \tag{5.4}$$

By Levy's inversion formula we have

$$\sum_{j=\ell+1}^{m} P(S_n=j) = \sum_{j=\ell+1}^{m} P(S_n'=j) + \P_1 + \P_2 + \P_3$$

where

$$|\P_1| \leq \frac{1}{2\pi}\int_{-\pi}^{\pi}\left|\frac{1}{2}\sum_{k=1}^{n}\left(\int_{-\infty}^{\infty} W\right)^2 e^{\sum_{k=1}^{n}\left[E\left(e^{itX_k}\right)-1\right]}\right|\frac{du}{|t|/\pi}$$

and $\P_2$ and $\P_3$ are the parts of R_2 and R_3, respectively, in Lévy's inversion formula. Using the following inequalities

$$\int_{-\infty}^{\infty} W = \left(\int_{-\infty}^{0-} + \int_{0}^{\infty}\right) W \leq t\int_{-\infty}^{\infty}|x|dF_k(x), \tag{5.5}$$

$$\left[\int_{\substack{-\infty\\x\neq 0}}^{\infty} xdF_k(x)\right]^2 \leq \left[\int_{-\infty}^{\infty} x^2dF_k(x)\right]\left[\int_{\substack{-\infty\\x\neq 0}}^{\infty} dF_k(x)\right] = \alpha_k\int_{\infty}^{\infty} x^2dF_x(x), \tag{5.6}$$

and

$$\left|e^{-\sum_{k=1}^{n}\int_{-\infty}^{\infty} W}\right| \leq e^{-\sum_{k=1}^{n}\int_{-\infty}^{\infty}(1-\cos tx)dF_k(x)} \leq e^{-\frac{2t^2}{\pi^2}\sum_{k=1}^{n}\int_{-\infty}^{\infty}x^2dF_k(x)},$$

we have

$$|\P_1| \leq \frac{1}{4\pi}\int_{-\pi}^{\pi} t^2 \cdot \sum_{k=1}^{n}\left[\int_{-\infty}^{\infty} xdF_k(x)\right]^2 \cdot e^{-\frac{2t^2}{\pi^2}\sum_{k=1}^{n}\int_{-\infty}^{\infty}x^2dF_k(x)}\frac{dt}{|t|/\pi}$$

$$\leq \frac{\tilde{\alpha}}{4}\int_{-\pi}^{\pi}\left[\sum_{k=1}^{n}\int_{-\infty}^{\infty} x^2dF_k(x)\right]|t|\, e^{-\frac{2t^2}{\pi^2}\sum_{k=1}^{n}\int_{-\infty}^{\infty}x^2dF_1(x)}\, dt$$

$$= \frac{\pi^2}{8}\tilde{\alpha}; \text{ and}$$

$$|\P_2| \leq \frac{1}{2\pi}\int_{-\pi}^{\pi}\frac{\tilde{\alpha}^2t^4}{2}\left[\sum_{k=1}^{n}\int_{-\infty}^{\infty} x^2dF_k(x)\right]^2 e^{\text{EXP}}\frac{dt}{|t|/\pi},$$

$$\text{where EXP} = \frac{\tilde{\alpha}t^2}{2}\sum_{k=1}^{n}\int_{-\infty}^{\infty} x^2dF_k(x) - \frac{2t^2}{2}\sum_{k=1}^{n}\int_{-\infty}^{\infty}x^2dF_k(x)$$

$$\leq \frac{\tilde{\alpha}^2}{2} \int_0^\infty t^3 \left[\sum_{k=1}^n \int_{-\infty}^\infty x^2 dF_k(x) \right]^2 e^{-t^2\left(\frac{2}{\pi^2} - \frac{\tilde{\alpha}}{2}\right) \sum_{k=1}^n \int_\infty^\infty x^2 dF_k(x)} dt$$

$$= \frac{\tilde{\alpha}^2}{4} \left(\frac{2}{\pi^2} - \frac{\tilde{\alpha}}{2} \right)^{-2} ;$$

$$|\P_3| \leq \frac{1}{2\pi} \int_{-\pi}^{\pi} \frac{2\tilde{\alpha}^2}{3(1-2\tilde{\alpha})} t^2 \cdot \left[\sum_{k=1}^n \int_{-\infty}^\infty x^2 dF_k(x) \right] \text{ mult. by}$$

$$\exp \left\{ \left[-\frac{2t^2}{\pi^2} + \frac{\tilde{\alpha} t^2}{2} + \frac{2\tilde{\alpha}^2}{3(1-2\tilde{\alpha})} \right] \left[\sum_{k=1}^n \int_{-\infty}^\infty x^2 dF_k(x) \right] \right\} \frac{dt}{|t|/\pi}$$

$$\leq \frac{\tilde{\alpha}^2}{3} (1-2\tilde{\alpha})^{-1} \left[\frac{2}{\pi^2} - \frac{\tilde{\alpha}}{2} - \frac{2\tilde{\alpha}^2}{3(1-2\tilde{\alpha})} \right]^{-1} ,$$

since

$$\left| \sum_{k=1}^n \int_0^{\int_{-\infty}^\infty W} \frac{z^2}{1-z} dz \right| \leq \frac{2\tilde{\alpha}^2}{3(1-2\tilde{\alpha})} t^2 \cdot \sum_{k=1}^n \int_{-\infty}^\infty x^2 dF_k(x).$$

REFERENCES

[1] Franken, Peter 1963/64. Approximation der Verteilungen von Summen unabhängiger nichtnegativer ganzzahliger Zufallsgrössen durch Poissonschen Verteilungen. Math. Nachr. 27:303-340.

[2] Govindarajulu, Zakkula 1963. Normal approximations to the classical discrete distributions. Proc. Internat. Symp. Classical and Contagious Discrete Distributions, Montreal, 79-108. Also 1965 Sankhyā Series A, 27:143-72.

[3] Herrmann, Horst 1965. Variationsabstand jurschen der Verteilung einer Summe unabhängiger nichtnegativer ganzzahliger Zufallsgrössen und Poissonschen Verteilungen. Math. Nachr. 29:265-89.

[4] LeCam, Lucien 1960. An approximation theorem for Poisson binomial distribution. Pacific Journ. Math. 10:1181-97.

[5] ———. 1963. A note on the distribution of sums of independent random variable. Proc. Nat. Acad. Sci., U.S.A., 50:601-3.

[6] Makabe, Hajime 1955. A normal approximation to binomial distribution. Rep. Stat. Appl. Res. JUSE 4:47-53.

[7] ————., and Morimura, H. 1956. On the approximations to some limiting distributions. Kōdai Math. Sem. Rep.

[8] ————. 1962. On the approximation to some limiting distributions with some applications. Kōdai Math. Sem. Rep. 14:123-33.

[9] ————. 1964. On the approximation to some limiting distributions with applications to the theory of sampling inspections by attributes. Kōdai Math. Sem. Rep. 16:1-17.

[10] Zolotarev, V. M. 1965. On the closeness of the distribution of two sums of independent random variables. Theory of Probability and its Applications. 10:472-9.

NOTE ON AN APPROXIMATION TO THE DISTRIBUTION OF THE RANGE FROM POISSON SAMPLES

B. M. BENNETT
Department of Preventive Medicine
University of Washington
Seattle, Washington

E. NAKAMURA
Veterans Administration Hospital
Sepulveda, California

SUMMARY

An approximation to the distribution of the range for the symmetric multinomial (Johnson and Young, 1960; Young, 1962) is compared with the exact percentage points of the symmetric multinomial (Bennett and Nakamura, 1968). The use of this approximation has been suggested by Pettigrew and Mohler (1967) for the corresponding conditional distribution of the sample Poisson range.

NOTE ON AN APPROXIMATION TO THE DISTRIBUTION OF THE RANGE FROM POISSON SAMPLES

If $x_1, x_2, \ldots, x_n$ represent a series of discrete observations ($\Sigma x_i = x$), the sample range r has been proposed as a test of homogeneity of the observations from an equiprobable or symmetric multinomial distribution (see [2] and [5]). Equivalently r may be used in testing the agreement of n independent observations from a common Poisson distribution

$$e^{-\lambda}(\lambda^{x_i}/x_i!),\ x_i = 0,1,2,\ldots, \tag{1}$$

with parameter λ, since the conditional distribution of the x_i's, for x = constant, is also that of a symmetric multinomial with equal probabilities (=1/n).

As a rapid test of agreement of n sample observations with (1), [4] suggests the use of one of the approximations in [2], $r' \doteqdot \sqrt{\bar{x}}w$; i.e.,

$$P\{r \leq r_\alpha\} \doteqdot P\{\sqrt{\bar{x}}\ w \leq \sqrt{\bar{x}}W_\alpha\} \tag{2}$$

In (2) w represents the sample range from n independent unit normal variates, and $n\bar{x} = \Sigma x_i = x$. Selected upper and lower percentage points W_α of the cumulative distribution of w have been extensively tabulated by Owen [3], Table 6, for various α and $n \leq 100$. Also [4] reproduces some of the values W_α for n = 1(1)10, and suggests a graphical procedure using the approximation (2) to examine a series of n observations with respect to an assumed distribution (1).

Recent and extensive calculations (in [1]) of the upper percentage points of r from a symmetric multinomial, or values r_α such that

$$P\{r \geq r_\alpha\} = \sum_{r \geq r_\alpha} \frac{x!}{\Pi\, x_i!}\left(\frac{1}{n}\right)^x \leq \alpha, \tag{3}$$

and also $P\{r+1 > r_\alpha\}$, for α = .05, .01, n = 2(1)10, permits a detailed examination of the accuracy of the approximation r' from (2). Since there may also be occasional interest in the <u>under</u>-dispersion of a particular sample, the lower 5% and 1% points have also been compared.

For selected values of n and sample mean $\bar{x}$, the following Tables 1-4 compare the approximation r' with the nominal values r_α from (3). Also given are the exact significance probabilities

$P\{r \geq r_\alpha\}$. Since these latter probabilities differ frequently from α, the corresponding probabilities associated with the next smaller integer ($=r_\alpha-1$) are given for the case $n > 2$. For the special case, $n = 2$, the corresponding integer is $r_\alpha-2$.

In particular it will be noted that the approximation r' (when rounded up to the next integer) usually agrees closely with the corresponding r-value for the upper percentage points. As n becomes large the approximation is close even for small values of $\bar{x}$.

Young [5] has also suggested a continuity correction for the approximation to the multinomial range of the form $r'' = r' + \delta$, where $\delta = 1$ for $n = 2$ and $\delta = 1/2$ for $n > 2$. Although [4] does not mention the use of this correction, detailed comparisons of the exact significance levels of r" with those of r' in Tables 1-4 on pages 122-25 indicated that the use of r" is unnecessary.

REFERENCES

[1] Bennett, B. M., and Nakamura, E. 1968. Percentage points of the range from a symmetric multinomial distribution. Biometrika 55:377-379.

[2] Johnson N. L. and Young, D. H. 1960. Some applications of two approximations to the multinomial distribution. Biometrika 47:460-469.

[3] Owen, D. B. 1962. Handbook of Statistical Tables. Addison Wesley, Reading, Mass.

[4] Pettigrew, H. M. and Mohler, W. C. 1967. A rapid test for the Poisson distribution using the range. Biometrics 23 685-692.

[5] Young, D. H. 1962. Two alternatives to the standard X^2-test of the hypothesis of equal frequencies. Biometrika 49: 107-116.

Table 1. Comparisons of Lower Percentage Points of r, r'

		α = 0.99				α = 0.95			
n	$\bar{x}$	$r_{.99}$	$P\{r \geq r_{.99}\}$	$P\{r \geq r_{.99}-2\}$	$r'_{.99}$	$r_{.95}$	$P\{r \geq r_{.95}\}$	$P\{r \geq r_{.95}-2\}$	$r'_{.95}$
2	5	2	.754	1.000	.09	2	.754	1.000	.24
	10	2	.824	1.000	.10	2	.824	1.000	.33
	15	2	.856	1.000	.11	2	.856	1.000	.39
	20	2	.875	1.000	.13	2	.875	1.000	.44
n	x	$r_{.99}$	$P\{r \geq r_{.99}\}$	$P\{r \geq r_{.99}-1\}$	$r'_{.99}$	$r_{.95}$	$P\{r \geq r_{.95}\}$	$P\{r \geq r_{.95}-1\}$	$r'_{.95}$
3	1	-	--	--	0.2	-	--	--	0.6
	2	2	.877	1.000	0.3	2	.877	1.000	0.3
	5	2	.947	1.000	0.4	2	.947	1.000	1.0
	10	2	.973	1.000	0.6	3	.826	.973	1.4
4	1	2	.906	1.000	0.4	2	.906	1.000	0.8
	2	2	.962	1.000	0.6	3	.551	.962	1.1
	5	2	.989	1.000	1.0	3	.838	.989	1.7
	10	3	.934	.996	1.4	3	.934	.996	2.4

Table 2. Comparisons of Lower Percentage Points of r, r'

n	$\bar{x}$	$r_{.99}$	$P\{r \geq r_{.99}\}$	$P\{r \geq r_{.99}-1\}$	$r'_{.99}$	$r_{.95}$	$P\{r \geq r_{.95}\}$	$P\{r \geq r_{.95}-1\}$	$r'_{.95}$
5	1	2	.962	1.000	0.7	3	.290	.962	1.0
	2	3	.988	1.000	0.9	4	.679	.988	1.5
	5	3	.919	.998	1.5	3	.919	.998	2.3
6	1	2	.985	1.000	0.9	3	.367	.985	1.3
	2	3	.770	.997	1.2	3	.770	.997	1.8
	5	3	.960	.9996	1.9	4	.882	.960	2.8
8	1	3	.499	.998	1.2	3	.499	.998	1.6
	2	3	.882	.9997	1.7	3	.882	.9997	2.3
	3	3	.958	.9999	2.1	4	.848	.958	2.8
10	1	3	.604	.9996	1.5	3	.604	.9996	1.9
	2	3	.940	.9999	2.1	3	.940	.9999	2.6
	3	3	.984	.9999	2.5	4	.921	.984	3.2

Table 3. Comparisons of Upper Percentage Points of r, r'

			$\alpha = 0.05$				$\alpha = 0.01$		
n	$\bar{x}$	$r_{.05}$	$P\{r \geq r_{.05}\}$	$P\{r \geq r_{.05}-2\}$	$r'_{.05}$	$r_{.01}$	$P\{r \geq r_{.01}\}$	$P\{r \geq r_{.01}-2\}$	$r'_{.01}$
2	5	8	.021	.109	6.2	10	.002	.021	8.2
	10	10	.041	.115	8.8	14	.003	.012	11.6
	15	12	.043	.099	10.7	16	.005	.016	14.1
	20	14	.038	.081	12.4	18	.006	.017	16.3

n	$\bar{x}$	$r_{.05}$	$P\{r \geq r_{.05}\}$	$P\{r \geq r_{.05}-1\}$	$r'_{.05}$	$r_{.01}$	$P\{r \geq r_{.01}\}$	$P\{r \geq r_{.01}-1\}$	$r'_{.01}$
3	2	6	.004	.053	4.7	6	.004	.053	5.8
	5	8	.043	.093	7.4	10	.005	.015	9.2
	10	11	.048	.082	10.5	14	.006	.013	13.0
	15	14	.035	.056	12.8	17	.007	.012	16.0
4	1	4	.016	.203	3.6	-	--	--	4.4
	2	6	.017	.089	5.1	7	.002	.017	6.2
	5	9	.031	.075	8.1	11	.003	.011	9.8
	10	12	.047	.084	11.5	15	.005	.012	13.9

Table 4. Comparisons of Upper Percentage Points of r, r'

n	$\bar{x}$	$r_{.05}$	$P\{r \geq r_{.05}\}$	$P\{r \geq r_{.05}-1\}$	$r'_{.05}$	$r_{.01}$	$P\{r \geq r_{.01}\}$	$P\{r \geq r_{.01}-1\}$	$r'_{.01}$
5	1	4	.034	.290	3.9	5	.002	.034	4.6
	2	6	.029	.133	5.5	7	.004	.029	6.5
	5	9	.048	.113	8.6	11	.006	.018	10.3
6	1	5	.004	.052	4.0	5	.004	.052	4.8
	2	6	.043	.179	5.7	7	.007	.043	6.7
	5	10	.026	.068	9.0	11	.009	.026	10.6
8	1	5	.010	.090	4.3	5	.010	.090	5.0
	2	7	.015	.073	6.1	8	.002	.015	7.1
	3	8	.034	.112	7.4	9	.009	.034	8.6
10	1	5	.016	.127	4.5	6	.001	.016	5.2
	2	7	.023	.104	6.3	8	.004	.023	7.3
	3	9	.014	.051	7.7	10	.003	.014	8.9

PROPERTIES OF THE MAXIMUM LIKELIHOOD ESTIMATOR FOR THE PARAMETER OF THE LOGARITHMIC SERIES DISTRIBUTION

K. O. BOWMAN*
Computing Technology Center
Oak Ridge, Tennessee

L. R. SHENTON**
University of Georgia
Consultant, Oak Ridge National Laboratory

SUMMARY

This selection is concerned with extending previous discussions on the properties of the maximum likelihood estimator for a parameter of the logarithmic series distribution, which has many applications in scientific pursuits.

* Research sponsored by the U. S. Atomic Energy Commission under contract with the Union Carbide Corporation.

** Research sponsored by the Southeastern Forest Experiment Station with U. S. Department of Defense Funds, IPR-19-8-8001.

1. INTRODUCTION

The logarithmic series distribution has many applications in ecology as a model for the distribution of species abundance and in meterology as a model for the duration (in days or other units) of rainfall storms (in this connection see also the article by Shenton and Skees in this series. An extensive bibliography is given in [3].

Here we are concerned with properties of the maximum likelihood estimator $\hat{\theta}$ of the ratio parameter θ. Until quite recently, knowledge of the distribution of $\hat{\theta}$ was confined to its asymptotic variance and its asymptotic distribution. In 1964, Patil, Kamat, and Wani used expressions described by the present writers in [4] to evaluate the n^{-1} and n^{-2} coefficients in the bias and variance. We have now found a method of extending this procedure; and in certain special classes of estimator equations, terms of order up to n^{-9} can be evaluated, although there may be overflow or round-off errors locally in the parameter space. Another example is given in "Maximum likelihood estimator moments for the two-parameter gamma distribution" [6].

2. BASIC FORMULAE

If X is a logarithmic series random variate, then its probability function can be written

$$\Pr(X=x) = \alpha\theta^x/x, \tag{1}$$

$$\text{where } x = 1,2,\ldots;\ \alpha^{-1} = \ln 1/(1-\theta)\ ,\ 0 \leq \theta < 1,$$

and the maximum likelihood estimator of θ is a solution of the equation

$$\frac{\hat{\theta}}{(1-\hat{\theta})\ln(1/(1-\hat{\theta}))} = \bar{x} \tag{2}$$

where $\bar{x}$ is the mean of the random sample $x_1, x_2, \ldots x_n$. The solution $\hat{\theta}$ is unique, and in particular when $\bar{x} = 1$, $\hat{\theta} = 0$.

The stochastic Taylor expansion for $\hat{\theta}$ in terms of the random incremental variables $x_1 = x - E(x)$, where E refers to the usual expectation operator, is

$$\hat{\theta} = \theta + \sum_{r=1}^{\infty} \frac{x_1^r \bar{\theta}_r}{r!}, \tag{3}$$

where

$$\bar{\theta}_r = \left.\frac{d^r\hat{\theta}}{d\bar{x}^r}\right| \hat{\theta} = \theta,\ \bar{x} = \mu$$

$$\mu = E\bar{x} = \frac{-\theta}{(1-\theta)\ell n(1-\theta)} .$$

The convergence of (3), or the moment series derived from it by taking expectations of its powers, poses very difficult problems which we are unable to resolve. At least we should need the asymptotic form for $\bar{\theta}_r$ and also $E(\bar{x}_1)^r$ for regions of the parameter space (θ,n), and both of these are out of reach at present for want of tractable closed forms.

3. MOMENTS OF THE MEAN

We require the central moments (ν_s) of the mean $\bar{x}$ and we put

$$\nu_s = E(x-E\bar{x})^s \tag{4a}$$

$$\mu_s = E(x-\mu)^s, \text{ where } \mu=Ex. \tag{4b}$$

First of all we assume μ_s is known. Then the values of ν_s can be set up recursively from the relation

$$\sum_{r=0}^{s} \binom{s}{r} \nu_{s+1-r}\mu_r/n^r = \mu_{s+1}/n^s, \quad \text{where } s=1,2,\ldots,\ \nu_1=0. \tag{5}$$

This expression can be constructed following the lines of the development of formulae (11) and (13) given in [5], intended for use on a digital computer. As examples, we have

$$s = 1;\ \nu_2+\nu_1\mu_1/n = \mu_2/n$$

$$s = 2;\ \nu_3+2\nu_2\mu_1/n+\nu_1\mu_2/n^2 = \mu_3/n^2$$

so that the coefficients of powers of n^{-1} have to be isolated in using (5).

4. CENTRAL MOMENTS OF μ_s

[2] Introduces polynomials $C_n(\theta)$ defined by

$$C_n(\theta) = \sum_{k=0}^{n} c(n,k)\theta^k, \text{ where } n=0,1,\ldots \tag{6a}$$

$$c(n,k) = (k+1)c(n-1,k)+(n-k+1)c(n-1,k-1) \tag{6b}$$

$$c(0,0) = 1; \; c(0,k) = 0 \text{ for } k > 0.$$

Then the crude moments μ'_s have the form

$$\mu'_s = \frac{\alpha\theta}{(1-\theta)^s} C_{s-2}(\theta), \; (s = 2,3,\ldots), \tag{6c}$$

from which the central moments are readily derived using the expansion (4b). We used a computerized version of this approach to derive μ_s; finally (5) leads to the desired moments of the mean. We remark in passing that the computer output shows that the coefficients of powers of n^{-1} in the higher moments (as high as order eighteen for powers of n^{-1} to n^{-9}) are numerically large when θ is small and only moderately large when θ is near to unity.

Another method of deriving μ_s is worth noting because of its simplicity. Starting with the probability-generating function for the logarithmic series random variate X, namely

$$G(t) = \ln(1-\theta t)/\ln(1-\theta), \tag{7a}$$

we have for the generating function of the central moments

$$G^*(t) = E[\exp t(X-\mu)]$$

$$= [\exp(-\mu t)]\ln(1-\theta e^t)\ln(1-\theta), \tag{7b}$$

$$= \sum_0^\infty \mu_s t^s/s! \tag{7c}$$

But if

$$-\ln 1-\theta e^t = \sum_0^\infty C_s^* t^s/s!, \tag{8}$$

then by differentiation with respect to t and θ, we find, after simplification,

$$(1-\theta e^t)^{-1} = 1+C_1^*+C_2^*t/1!+C_3^*t^2/2!+\ldots \tag{9a}$$

$$= 1+\theta \sum_{s=0}^\infty \frac{t^s}{s!}\frac{d\theta_s^*}{d\theta}. \tag{9b}$$

Thus from (9a)

$$C_s^* = \theta\left\{1+\sum_{r=0}^{s-1}\binom{s-1}{r}C_{r+1}^*\right\} \; (s=1,2,\ldots). \tag{10a}$$

For example,

$$
\begin{aligned}
C_1^* &= \frac{\theta}{1-\theta},\\
C_2^* &= \frac{\theta}{(1-\theta)^2},\\
C_3^* &= \frac{\theta+\theta^2}{(1-\theta)^3},\\
C_4^* &= \frac{\theta+4\theta^2+\theta^3}{(1-\theta)^4},\\
C_5^* &= \frac{\theta+11\theta^2+11\theta^3+\theta^4}{(1-\theta)^5},
\end{aligned}
\tag{11}
$$

and so on. These are obviously related to the polynomials given in Table 1 by [2]. But from (9b) there is the simple relation

$$C_{s+1}^* = \theta\frac{dC_s^*}{d\theta},\ \text{where } s = 0,1,\ldots \text{ and} \tag{12}$$

where $C_0^* = -\ln(1-\theta)$. Hence for the central moments

$$\mu_s = \frac{1}{C_0^*}\left\{C_s^*-\binom{s}{1}\mu C_{s-1}^*+\binom{s}{2}\mu^2 C_{s-2}^*+\ldots+(-1)^s C_0^*\mu^s\right\}. \tag{13}$$

This leads to the compact formula

$$\mu_s = \frac{1}{\ln(1-\theta)}\left(\theta\frac{d}{d\theta}-\mu\right)^s \ln(1-\theta),\ \text{when } s = 0,1,2,\ldots \tag{14}$$

for the s-th central moment, μ being constant as far as the operator $d/d\theta$ is concerned; and $(\theta d/d\theta)^s$ is interpreted as a product of s factors $\theta d/d\theta$ and not $\theta^s d^s/d\theta^s$.

5. THE DERIVATIVES $\bar{\theta}_r$

Let $f(\theta) = -(1-\theta)\ln(1-\theta)/\theta$ (15)

so that the equation for the maximum likelihood estimate $\hat{\theta}$ is

$$f(\hat{\theta}) = \frac{1}{\bar{x}},$$

with derivative of order r

$$\frac{d^r f(\hat{\theta})}{d\bar{x}^r} = \frac{(-1)^r r!}{\bar{x}^{r+1}} \tag{16}$$

But writing $f^{(m)}(\theta)$ for $d^m f(\theta)/d\theta^m$ and using Faà di Bruno's formula (see [1]), we can find the derivatives $\bar{\theta}_r$ (knowing $f^m(\bar{\theta})$).

from

$$\sum_{m=0}^{r} \sum \left(\frac{\bar{\theta}_{p_1}}{p_1!}\right)^{\pi_1} \left(\frac{\bar{\theta}_{p_2}}{p_2!}\right)^{\pi_2} \cdots \left(\frac{\bar{\theta}_{p_m}}{p_m!}\right)^{\pi_m} \frac{r!f^{(\Sigma)}(\bar{\theta})}{\pi_1!\ \pi_2!\ldots\pi_m!}$$

$$= \frac{(-1)^r r!}{\bar{x}^{r+1}}, \qquad (17)$$

where the second summation is over all non-negative integer solutions to

$$p_1\pi_1+p_2\pi_2+\ldots+p_m\pi_m = r$$

and

$$\Sigma = \sum_{s=1}^{m} \pi_s .$$

To use this formula we need to know the derivatives $f^{(m)}(\bar{\theta})$. These are given recursively by the formula (obtained by repeated differentiation of (15)):

$$y(1-y)f^{m+1}(y)+[1+m(1-y)]f^m(y)-m(m-1)f^{m-1}(y)$$

$$= -\delta_{m,1}+(1-y)\delta_{m,0}, \text{ where} \qquad (18)$$

$m=0,1,\ldots$; $y=\bar{\theta}$, $f^0(y)=f(y)$, and $\delta_{r,s}$ is the Kronecker delta function. A program has been set up to evaluate $\bar{\theta}_1,\bar{\theta}_2\ldots$, using (17) and (18). In contrast to the moments ν_s, the derivatives $\bar{\theta}_r$ are large numerically for θ near to unity, and relatively small for θ near to zero.

6. COMMENTS ON THE MOMENTS

Coefficients of powers of n^{-1} up to n^{-9} for the first four moments of $\hat{\theta}$ are given in Table 1 and are graphically displayed in Figures 1-4. Convergence is most rapid for μ_1 and becomes progressively retarded for μ_2,μ_3, and μ_4. Moreover, as θ increases towards unity, the series for each moment becomes less stable; this is brought out by noticing the series for $\theta = 0.99$, $n = 9$, $n = 10$, and particularly for μ_4. In fact, it is clear that if we take n large enough to damp off the terms in the series for μ_4, then n will also be large enough to damp off the terms in the lower moments. As an overall remark, convergence (using the word in a broad sense) appears to be satisfactory for $\theta \leq 0.9$

and $n \geq 9$ (the case $0.9 < \theta < 1.0$ would need larger values of $n \geq$ for stabilization, particularly as $\theta \to 1$). This can be seen from the terms in the fourth moment for $\theta = 0.9$, which are:

Terms in Fourth Moment

$\theta = 0.9$

n	8	9	10	12
n^{-2}	.000,054	.000,048	.000,035	.000,024
n^{-3}	.000,287	.000,227	.000,147	.000,085
n^{-4}	.000,623	.000,438	.000,255	.000,122
n^{-5}	.000,705	.000,440	.000,231	.000,093
n^{-6}	.000,276	.000,153	.000,072	.000,024
n^{-7}	-.000,385	-.000,190	-.000,081	-.000,023
n^{-8}	-.000,588	-.000,258	-.000,099	-.000,023
n^{-9}	-.000,017	-.000,007	-.000,002	.000,000

Graphs of the bias and standard deviation of $\hat{\theta}$ for n = 9 (1) 20, 22, 25 (5) 50 (10) 100 are given in Figures 5 and 6. For many practical purposes these, along with linear interpolation, should give sufficient accuracy.

7. SKEWNESS AND KURTOSIS FOR THE DISTRIBUTION OF θ

The values of $\sqrt{\beta_1} = \mu_3/\sqrt{\mu_2^3}$, $\beta_2 = \mu_4/\mu_2^2$ for the distribution of $\hat{\theta}$ for values of $\theta = 0.1(0.1)0.9$ and n = 8,9,10,15,25,50,100 are given in Table 2. As might be expected, there is the greatest departure from normality for n small and θ nearly unity; actually, $\sqrt{\beta_1}$ changes from positive to negative as θ increases through $\theta \approx 0.35$, depending on the sample size. However, very marked departures from normality would only occur for $\theta > 0.9$ and for n moderately large or small, and this would correspond to a situation in which small samples were drawn from a long-tailed distribution.

A graphical display of the skewness and kurtosis is given in Figures 7-9.

8. VERIFICATION OF THE RESULTS

The moments and moment parameters $\sqrt{\beta_1}$, β_2 can be evaluated by a sample configuration method, provided the distribution sampled has a short tail (θ small). This has been done on a computer for $\theta = 0.1$, truncating the series at $x = 9$ and carrying computation to 16 digits. All sample configurations are set up; and for each, the value of $\hat{\theta}$ is derived by an iterative procedure using (2) in the form

$$f^*(\hat{\theta}) = \frac{-(1-\hat{\theta})}{\hat{\theta}} \ln(1-\hat{\theta}) - \frac{1}{\bar{x}}$$

with a starting value $\hat{\theta} = \frac{9\bar{x}-6-\sqrt{9\bar{x}^2-12\bar{x}+6}}{6\bar{x}-2}$ (derived from a rational fraction approximation to the logarithmic series probability function), taking $\hat{\theta} = 0$ whenever $\bar{x} = 1$. The results for samples $n = 2(1)10$ are given in Table 3. Agreement with the results derived from the asymptotic series was almost perfect for $n = 8,9,10$; larger values of n are difficult to handle from the multinomial point of view because of the large number of sample configurations involved.

Lower order moments for $n < 8$ can also be checked against the asymptotic series (Table 1c, with $n = 10$, $\theta = 0.1$, can be used). Thus, for $n = 5$, the asymptotic series leads to $\mu_1 = 0.075631$ in good agreement with the multinomial value $\mu_{1_2} = 0.07564$.

In addition, we remark that our n^{-1} and n^{-2} terms in the mean and variance agree (apart from a difference in sign in the bias, which we feel is a slip, since our results are confirmed otherwise) to within a small percentage with those given in [2].

As for sheer numerical accuracy in setting up the coefficients in the moments using (3), (5), (6), (17) and (18), we carried out a section of the computations on an IBM 360-75 (carrying 16 significant digits) and also on a CDC 1604A (24 significant digits). The printout on the former gave six digits and on the latter, seven digits. For $\theta \geq 0.1$ the agreement in all terms was nearly perfect. A selection of comparisons is given in Table 4.

Discrepancies do occur for small values of θ ($\theta < 0.1$) and particularly for the lower order moments. For example, the coefficients of the n^{-1} through n^{-9} terms in the first two moments for the IBM and CDC machines were as follows:

$\theta = 0.05$

Coefficient	Mean Value of $\hat{\theta}$ IBM	Mean Value of $\hat{\theta}$ CDC	Variance of $\hat{\theta}$ IBM	Variance of $\hat{\theta}$ CDC
n^{-1}	-8.06632(-02)	-8.06632(-02)	9.179965(-02)	9.179996(-02)
n^{-2}	1.165198(-01)	1.16520(-01)	-2.73259(-01)	-2.73259(-01)
n^{-3}	-1.510194(-01)	-1.51019(-01)	5.171138(-01)	5.17114(-01)
n^{-4}	1.646560(-01)	1.64656(-01)	-6.08932(-01)	-6.08932(-01)
n^{-5}	-1.087255(-01)	-1.08734(-01)	-1.886038(-01)	-1.88603(-01)
n^{-6}	-1.249536(-01)	-1.26906(-01)	3.784847(00)	3.78504(00)
n^{-7}	7.440283(-01)	2.08863(-01)	-1.382251(01)	-1.37693(01)
n^{-8}	-2.040141(00)	-1.69692(02)	3.265552(01)	4.93348(01)
n^{-9}	4.00048(00)	-5.90814(04)	-3.835869(01)	5.84326(03)

It will be seen that for the mean, agreement is good up to the n^{-4} term and is almost completely lost from the n^{-7} term onwards. The agreement for the variance is a little better, but is completely lost at the n^{-9} term. Agreement for the third and fourth moments improves, and for the latter there is agreement in the first digit for the highest coefficient. A possible explanation for the improved accuracy for the higher moments relates to the fact that the first moment involves an eighteenth derivative (and moment), the second moment involves a seventeenth derivative (and moment), and so on. These high-order terms only affect the high-order terms in n^{-1}; obviously, because of the recursive schemes involved in the moments (doubly interlaced) and derivatives (Faà di Bruno's formula plus a second order recurrence), higher-order terms involve many more sums of products than the lower-order terms. It is possible then that the inaccuracies appear through large summations of terms involving products of numbers of widely different orders of magnitude.

9. CONCLUSION AND DISCUSSION

We have derived moment parameters for the distribution of $\hat{\theta}$, the maximum likelihood estimator of θ in the logarithmic series distribution. The mean, variance, skewness ($\sqrt{\beta_1}$) and kurtosis (β_2) are given for $0.1 \leq \theta \leq 0.9$ and $n \geq 8$; smaller values of θ could be dealt with using the sample configuration method provided n is not too large. For small θ and large n the asymptotic series approach could be used, although "round-off" errors and similar computer difficulties would need careful attention.

The interval $0.9 < \theta < 1.0$ presents serious difficulties for the asymptotic series approach, although moments could be assessed for large to very large values of n, the sample size.

As a general statement the distribution of $\hat{\theta}$ is not too far removed from the normal distribution; however, departures from normality become serious when θ exceeds about 0.9, when the sample size is less than about nine, and when both of these conditions are satisfied.

Verification of the results has been sought through (a) the use of two digital computers; (b) the use of a sample configuration method for small values of θ and small sample size.

REFERENCES

[1] di Bruno, Faà, 1876. Théorie des formes binaires. Libraine Brero, Turin

[2] Patil, G. P., Kamat, A. R., and Wani, J. K. 1966. Certain Studies on the Structure and Statistics of the Logarithmic Series Distribution and Related Tables. Tech. Rep. Aerospace Res. Lab., Wright-Patterson Air Force Base, p. 389.

[3] Patil, G. P., and Sharadchandra W. Joshi. 1968. A Dictionary and Bibliography of Discrete Distributions. Oliver and Boyd, London.

[4] Shenton, L. R., and Bowman, K. O. 1963. Higher Moments of a Maximum Likelihood Estimate. J.R.S.S. (B) 25:305.

[5] ________, ________, and Reinfelds, J. 1967. Sampling Moments of Moment and Maximum Likelihood Estimators for Discrete Distributions. Technical Report Number 17, Computer Center, University of Georgia; also abstract in I.S.I. Proceedings, XLII 1967,359-61.

[6] ________, ________, 1969. Maximum Likelihood Estimators for the 2-Parameter Gamma Distribution. Sankhya Vol. 31.

[7] ________, and Skees, P. 1969. Some Statistical Aspects of Amounts and Duration of Rainfall. Sampling and Other Aspects of Statistical Ecology. Pennsylvania State University Press, University Park (in preparation).

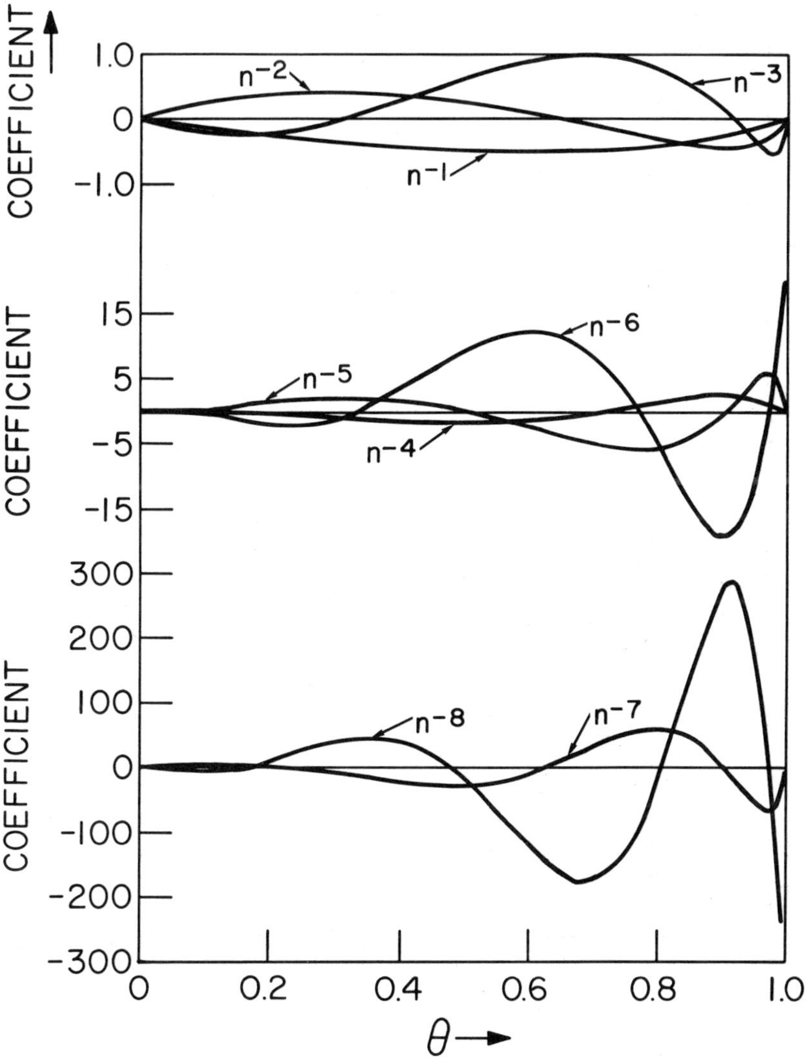

Figure 1. Logarithmic Series Distribution Maximum Likelihood Estimator $\hat{\theta}$ Bias n^{-1} thru n^{-8} Coefficients

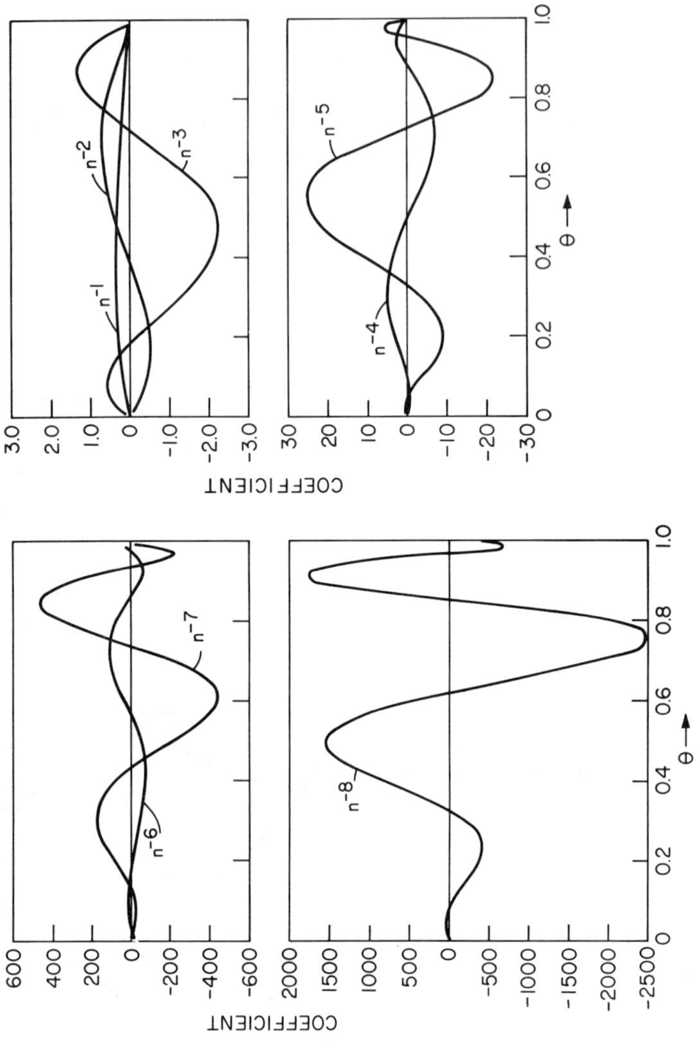

Figure 2. Logarithmic Series Distribution Maximum Likelihood Estimator $\hat{\theta}$
Var $\hat{\theta}$ n^{-1} thru n^{-8} Coefficients

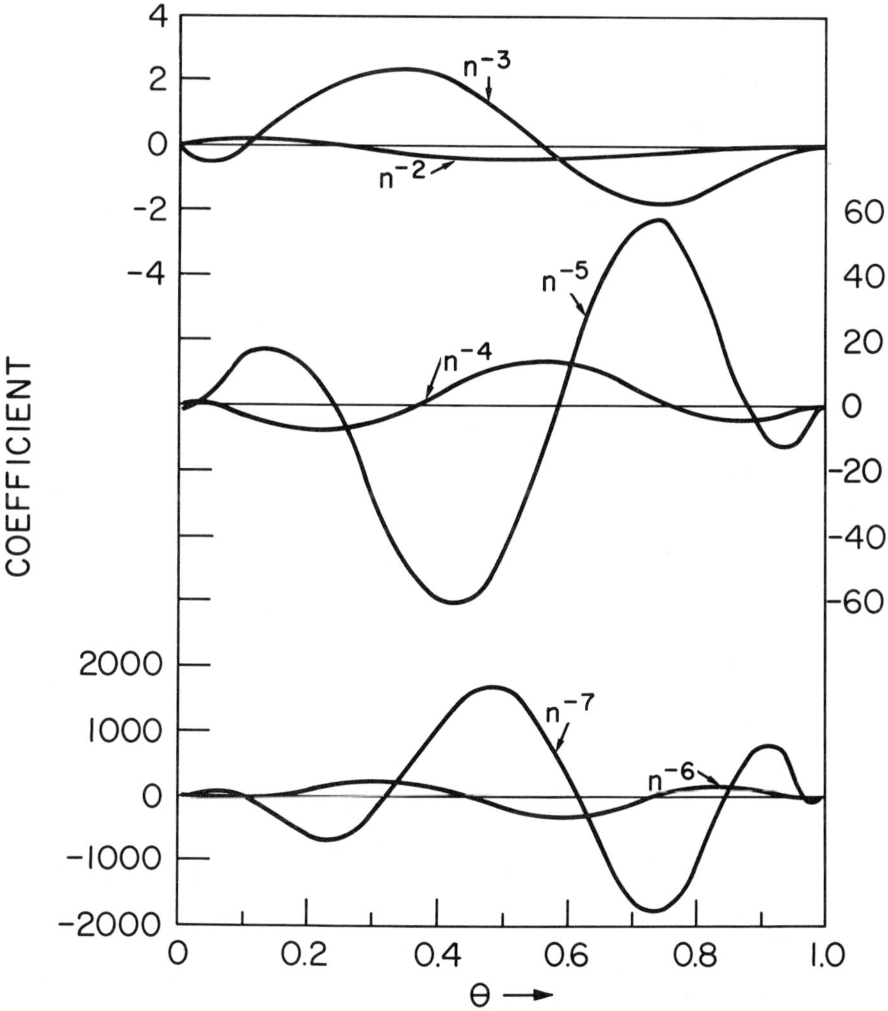

Figure 3. Logarithmic Series Distribution Maximum Likelihood Estimator $\hat{\theta}$ Third Moment $\mu_3(\hat{\theta})$, n^{-2} thru n^{-7} Coefficients

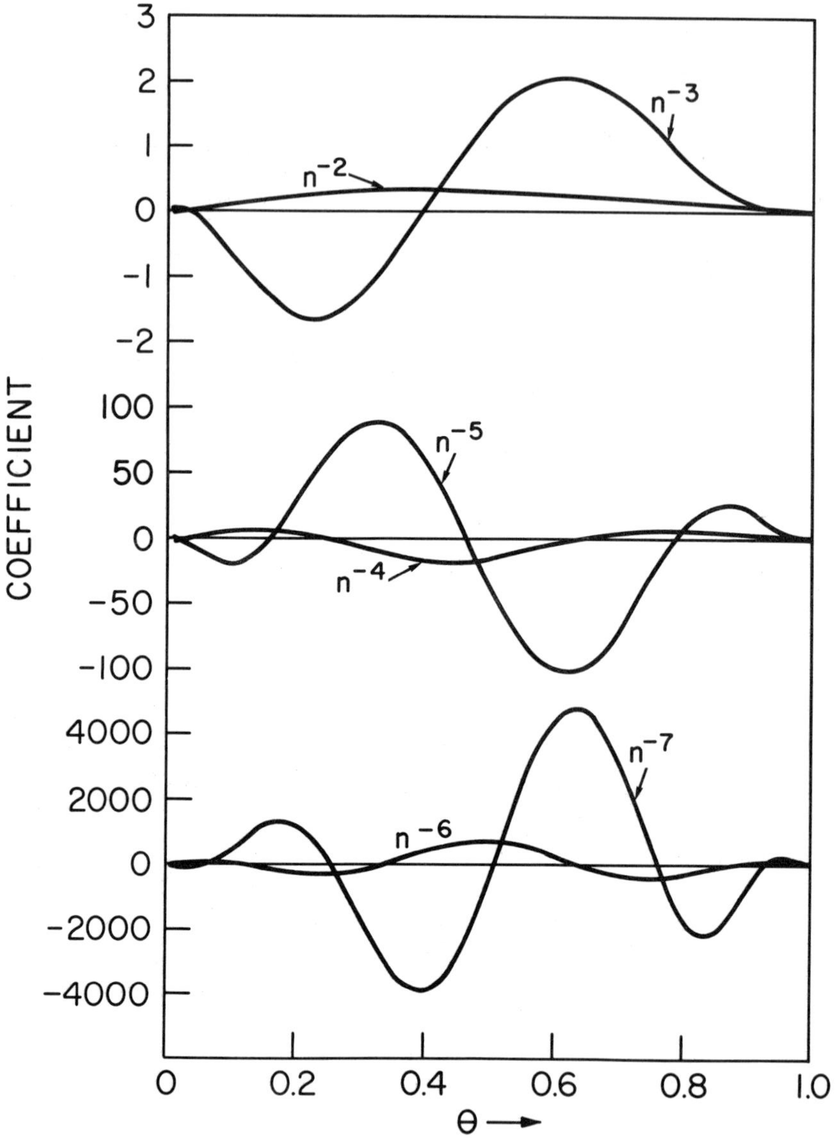

Figure 4. Logarithmic Series Distribution Maximum Likelihood Estimator $\hat{\theta}$ Fourth Moment $\mu_4(\hat{\theta})$ n^{-2} thru n^{-7} Coefficients

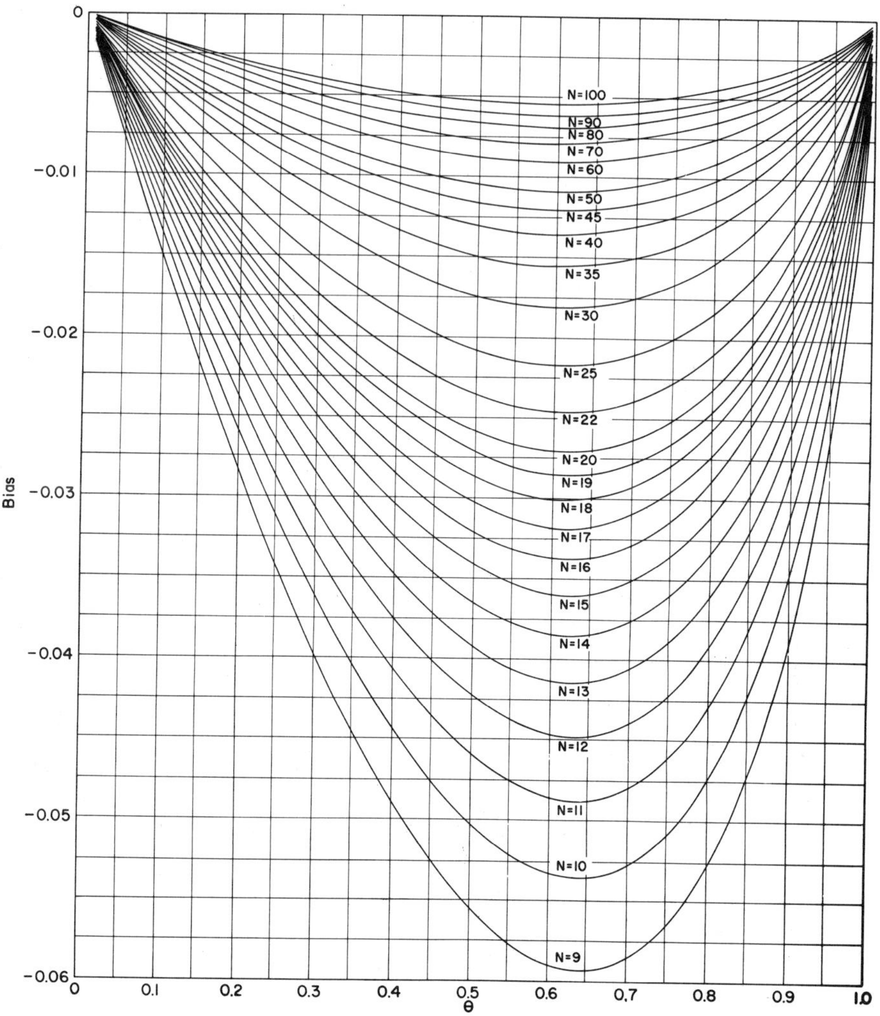

Figure 5. Logarithmic Series Distribution Bias of Maximum Likelihood Estimator $\hat{\theta}$

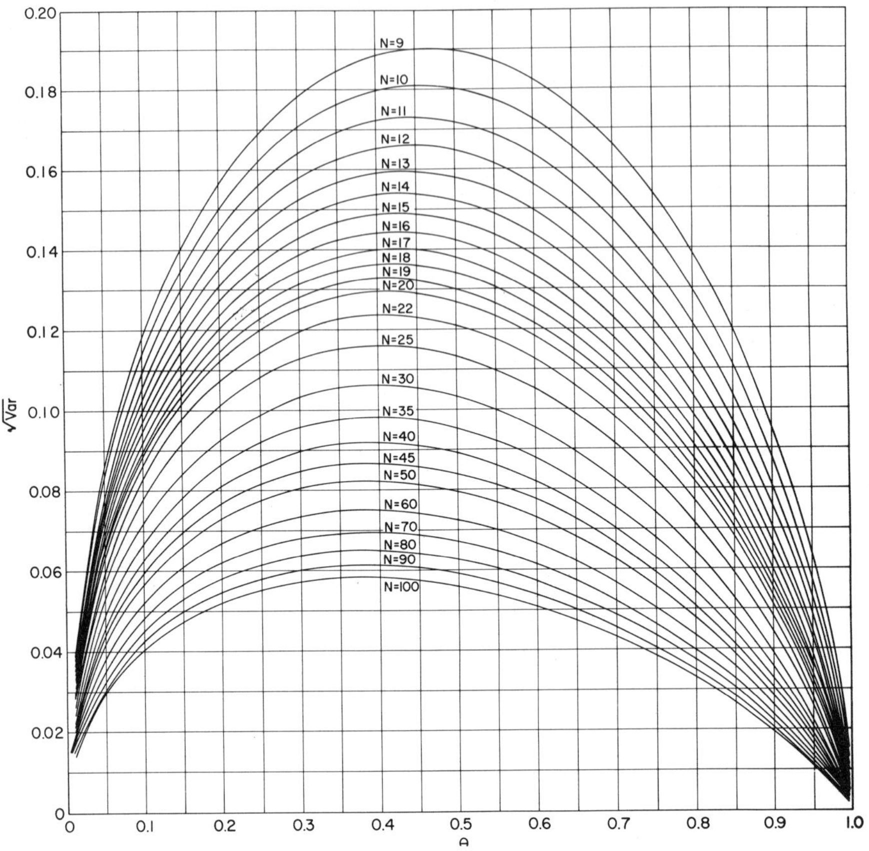

Figure 6. Logarithmic Series Distribution Standard Deviation of Maximum Likelihood Estimator $\hat{\theta}$

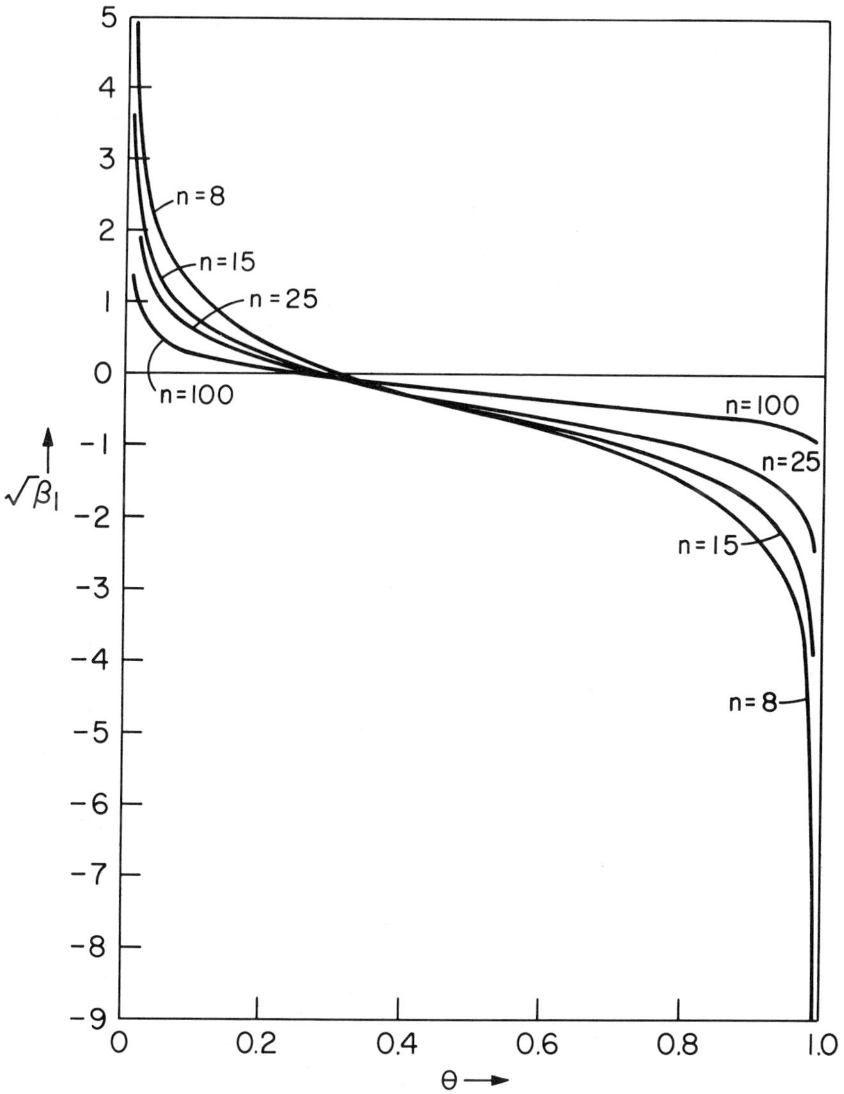

Figure 7. Logarithmic Series Distribution Skewness ($\sqrt{\beta_1}$) for Maximum Likelihood Estimator

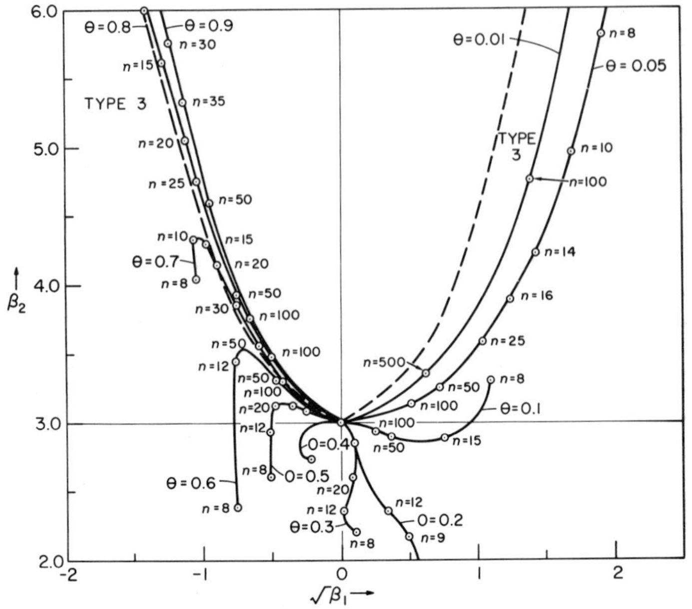

Figure 8. Logarithmic Series Distribution $\sqrt{\beta_1}(\hat{\theta})$, $\beta_2(\hat{\theta})$ for Maximum Likelihood Estimator

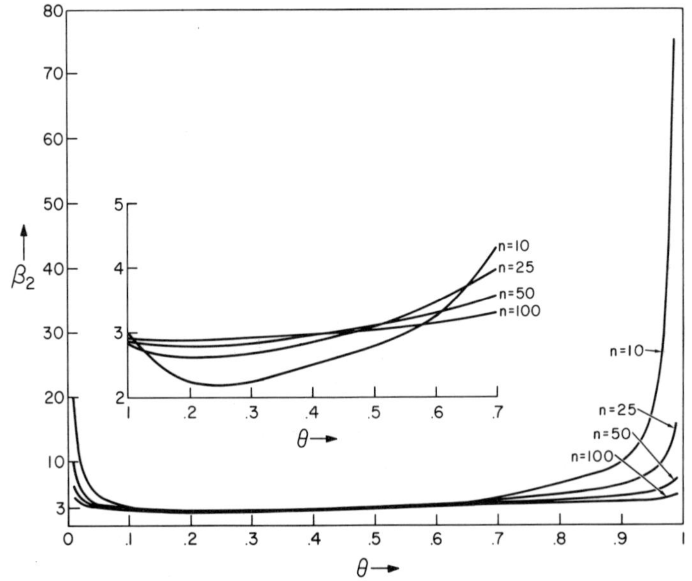

Figure 9. $\beta_2\ (\hat{\theta})$ for Logarithmic Distribution

Table 1a. Terms in Expansions in Powers of n^{-1} of Moments of $\hat{\theta}$
(All terms increased by a factor 10^6)

	$\theta = 0.1$ $n = 8$				$\theta = 0.5$ $n = 8$			
	μ_1	μ_2	μ_3	μ_4	μ_1	μ_2	μ_3	μ_4
n^{-1}	-19482	20967	——	——	-66203	38867	——	——
n^{-2}	3293	-6830	2762	1319	3477	5409	-7271	4532
n^{-3}	-460	1140	-297	-1252	1142	-4413	1902	2793
n^{-4}	39	46	-618	1162	-359	-40	2869	-3766
n^{-5}	5	-115	447	-626	2	718	-1345	-1065
n^{-6}	-4	48	-151	85	34	-192	-708	2811
n^{-7}	1	-11	2	171	-11	-104	795	-308
n^{-8}	0	-1	34	-183	-1	93	2	-1837
n^{-9}	0	2	-25	94	3	-7	-398	1095

	$\theta = 0.8$ $n = 8$				$\theta = 0.95$ $n = 8$			
	μ_1	μ_2	μ_3	μ_4	μ_1	μ_2	μ_3	μ_4
n^{-1}	-55863	12800	——	——	-23400	1302	——	——
n^{-2}	-5682	9612	-2655	492	-6913	1981	-120	51
n^{-3}	1505	1721	-3155	1692	-700	1547	-336	356
n^{-4}	313	-1273	-636	1951	384	623	-469	109
n^{-5}	-160	-513	1222	218	126	-54	-374	2040
n^{-6}	-22	300	646	-1437	-45	-217	-96	2440
n^{-7}	29	158	-529	-816	-27	-69	155	1488
n^{-8}	-1	-117	-442	-1011	9	70	185	-692
n^{-9}	-7	-57	322	1034	8	58	6	-2485

Table 1b. Terms in Expansions in Powers of n^{-1} of Moments of $\hat{\theta}$
(All Terms Increased by a Factor of 10^6)

	$\theta = 0.01$ $n = 9$				$\theta = 0.1$ $n = 9$			
	μ_1	μ_2	μ_3	μ_4	μ_1	μ_2	μ_3	μ_4
n^{-1}	-1840	2185	——	——	-17317	18638	——	——
n^{-2}	310	-792	461	14	2602	-5396	2183	1042
n^{-3}	-49	200	-241	78	-323	801	-209	-879
n^{-4}	7	-42	76	-50	24	29	-386	726
n^{-5}	-1	7	-17	14	3	-64	248	-347
n^{-6}	0	-1	2	0	-2	24	-75	42
n^{-7}	0	0	0	-2	0	-5	0	75
n^{-8}	0	0	0	2	0	0	13	-71
n^{-9}	0	0	0	0	0	0	-9	33

	$\theta = 0.4$ $n = 9$				$\theta = 0.7$ $n = 9$			
	μ_1	μ_2	μ_3	μ_4	μ_1	μ_2	μ_3	μ_4
n^{-1}	-53182	37673	——	——	-57978	20134	——	——
n^{-2}	4111	71	-4466	4258	-1904	8845	-4090	1216
n^{-3}	344	-2861	3001	12	1324	-543	-2269	2435
n^{-4}	-198	572	592	-2456	-18	-1115	947	852
n^{-5}	28	180	-999	1028	-86	108	898	-1202
n^{-6}	6	-135	198	729	16	198	-307	-745
n^{-7}	-4	18	213	-845	7	-49	-335	674
n^{-8}	0	19	-159	67	-4	-44	162	493
n^{-9}	0	-13	3	413	0	24	136	-495

	$\theta = 0.95$ $n = 9$				$\theta = 0.99$ $n = 9$			
	μ_1	μ_2	μ_3	μ_4	μ_1	μ_2	μ_3	μ_4
n^{-1}	-20800	1158	——	——	-5935	65	——	——
n^{-2}	-5462	1565	-95	4	-2612	133	-2	0.01
n^{-3}	-491	1087	-236	25	-770	160	-6	0.12
n^{-4}	240	389	-293	68	-68	134	-13	0.52
n^{-5}	70	-30	-208	113	59	78	-20	1.53
n^{-6}	-22	-107	-47	120	23	21	-23	3.38
n^{-7}	-12	-30	68	65	-6	-11	-19	5.93
n^{-8}	4	27	72	-27	-5	-16	-8	8.24
n^{-9}	3	20	2	-86	0	-4	4	8.56

Table 1c. Terms in Expansion in Powers of n^{-1} of Moments of $\hat{\theta}$
(All Terms Increased by Factor 10^6)

	θ = 0.01 n = 10				θ = 0.1 n = 10			
	μ_1	μ_2	μ_3	μ_4	μ_1	μ_2	μ_3	μ_4
n^{-1}	-1656	1967	——	——	-15586	16774	——	——
n^{-2}	251	-642	374	12	2107	-4371	1768	844
n^{-3}	-36	146	-175	57	-236	584	-152	-641
n^{-4}	5	-27	50	-33	16	19	-253	476
n^{-5}	-1	4	-10	9	2	-37	146	-21
n^{-6}	0	-1	1	0	0	13	-40	2
n^{-7}	0	0	0	0	0	-2	0	4
n^{-8}	0	0	0	0	0	0	1	-3
n^{-9}	0	0	0	0	0	0	0	1

	θ = 0.4 n = 10				θ = 0.7 n = 10			
	μ_1	μ_2	μ_3	μ_4	μ_1	μ_2	μ_3	μ_4
n^{-1}	-47864	33905	——	——	-52181	1812	——	——
n^{-2}	3330	57	-3617	3449	-1542	716	-3313	985
n^{-3}	251	-2086	2188	9	968	-40	-1654	1775
n^{-4}	-130	376	388	-1611	-12	-73	621	559
n^{-5}	17	106	-590	607	-51	6	530	-709
n^{-6}	3	-71	105	387	8	11	-163	-396
n^{-7}	-2	9	102	-404	3	-2	-160	322
n^{-8}	0	8	-68	29	-2	-2	70	212
n^{-9}	0	-5	1	160	0	1	53	-192

	θ = 0.95 n = 10				θ = 0.99 n = 10			
	μ_1	μ_2	μ_3	μ_4	μ_1	μ_2	μ_3	μ_4
n^{-1}	-18720	1042	——	——	-5341	58	——	——
n^{-2}	-4424	1268	-77	3	-2116	11	-1	0.01
n^{-3}	-358	792	-172	18	-561	116	-4	0.08
n^{-4}	157	255	-192	45	-45	88	-9	0.34
n^{-5}	41	-18	-123	67	35	46	-12	0.90
n^{-6}	-12	-57	-25	64	12	11	-12	1.80
n^{-7}	-6	-14	33	31	-3	-5	-9	2.83
n^{-8}	2	12	31	-12	-2	-7	-4	3.54
n^{-9}	1	8	8	-33	0	-2	2	3.32

Table 2. Skewness and Kurtosis

$\sqrt{\beta_1}$ and β_2 for Maximum Likelihood Estimator $\hat{\theta}$

θ/n		8	9	10	15	25	50	100
0.1	$\sqrt{\beta_1}$	1.14	1.06	1.00	0.77	0.57	0.38	0.26
	β_2	3.316	3.15	3.05	2.88	2.85	2.89	2.94
0.2	$\sqrt{\beta_1}$	0.48	0.43	0.39	0.26	0.16	0.09	0.05
	β_2	1.93	2.14	2.23	2.43	2.62	2.80	2.90
0.3	$\sqrt{\beta_1}$	0.10	0.06	0.03	-0.05	-0.09	-0.09	-0.07
	β_2	2.17	2.20	2.25	2.48	2.69	2.85	2.98
0.4	$\sqrt{\beta_1}$	-0.22	-0.24	-0.26	-0.29	-0.28	-0.23	-0.17
	β_2	2.73	2.54	2.52	2.69	2.87	2.96	2.99
0.5	$\sqrt{\beta_1}$	-0.51	-0.51	-0.52	-0.51	-0.45	-0.35	-0.26
	β_2	2.62	2.71	2.80	3.03	3.12	3.11	3.07
0.6	$\sqrt{\beta_1}$	-0.76	-0.77	-0.77	-0.73	-0.62	-0.47	-0.34
	β_2	2.37*	2.97	3.24	3.52	3.47	3.30	3.17
0.7	$\sqrt{\beta_1}$	-1.04	-1.06	-1.06	-0.97	-0.81	-0.59	-0.42
	β_2	4.04*	4.25	4.33	4.29	3.98	3.56	3.29
0.8	$\sqrt{\beta_1}$	-1.54	-1.50	-1.46	-1.28	-1.03	-0.74	-0.52
	β_2	8.02*	6.96	6.47	5.61	4.76	3.92	3.47
0.9	$\sqrt{\beta_1}$	-2.38*	-2.24	-2.14	-1.78	-1.36	-0.94	-0.65
	β_2	7.50*	8.94*	9.43**	8.48	6.34	4.58	3.76

* Not reliable because of slow rate of series convergence.

** Slightly suspect because of slow convergence.

Table 3. Moments of $\hat{\theta}$ using
(a) Multinomial Methods, (b) Asymptotic Series*

$\theta = 0.1$

Multonomial Probabilities

P_1 = .94912216	P_4 = .00023728	P_7 = .00000014
P_2 = .04745611	P_5 = .00001898	P_8 = .00000001
P_3 = .00316374	P_6 = .00000158	P_9 = .00000000

n	μ_1	μ_2	μ_3	μ_4	$\sqrt{\beta_1}$	β_2
2	0.05464	0.02743	0.01248	0.00659	2.748	8.757
3	0.06472	0.02532	0.00864	0.00381	2.145	5.949
4	0.07117	0.02273	0.00613	0.00242	1.790	4.690
5	0.07564	0.02038	0.00452	0.00167	1.552	4.023
6	0.07892	0.01837	0.00344	0.00123	1.381	3.631
7	0.08142	0.01668	0.00269	0.00094	1.251	3.386
8	0.08339	0.01525	0.00216	0.00075	1.148	3.225
	(0.08339)	(0.01525)	(0.00216)	(0.00075)	(1.148)	(3.225)
9	0.08499	0.01403	0.00177	0.00061	1.065	3.116
	(0.08499)	(0.01403)	(0.00177)	(0.00062)	(1.063)	(3.149)
10	0.086303	0.012978	0.001473	0.000,512	0.996	3.040
	(0.086303	(0.012978)	(0.001472)	(0.000,514)	(0.996)	(3.053)

* The entries in this table are those derived by a "multinomial" approach; those in parenthesis are derived from the asymptotic series.

Table 4. Comparison of Accuracy for Two Digital Machines

$\theta = 0.1$

Moment	Machine	Coefficient n^{-6}	n^{-7}	n^{-8}	n^{-9}
μ_1'	CDC	-9.802963-01	2.461617+00	-3.750214+00	-6.570077-01
	IBM	-9.80302-01	2.46081+00	-3.87420+00	-2.20258-01
μ_2	CDC	1.271217+01	-2.270206+01	-1.350728+01	3.013802+02
	IBM	1.27122+01	-2.27019+01	-1.34827+01	3.05616+02
μ_3	CDC	-3.971072+01	4.217957+00	5.728164+02	-3.398961+03
	IBM	-3.97107+01	4.21794+00	5.72813+02	-3.39953+03
μ_4	CDC	2.21946+01	3.590802+02	-3.063990+03	1.263924+04
	IBM	2.21943+01	3.59080+02	-3.06399+03	1.26393+04

$\theta = 0.2$

Moment	Machine	n^{-6}	n^{-7}	n^{-8}	n^{-9}
μ_1'	CDC	-2.526752+00	1.711469+00	1.073254+01	-5.704685+01
	IBM	-2.52675+00	1.71147+00	1.073254+01	-5.70520+01
μ_2	CDC	5.638359+00	6.903632+01	-3.643104+02	7.819357+02
	IBM	5.63836+00	6.90363+01	-3.64310+02	7.81938+02

UNBIASED ESTIMATION BASED ON INVERSE POISSON SAMPLING

MARTIN SANDELIUS
University of Gothenburg, Sweden

SUMMARY

Unbiased estimators of the Poisson mean and its reciprocal and unbiased estimators of the variances of these estimators are obtained for a sampling procedure which consists of taking observations until their sum attains or exceeds a given number.

1. INTRODUCTION

In three previous papers ([25], [26], and [27]) some inverse sampling procedures were proposed for the estimation of bacterial densities from plate counts. One of the procedures (see [26]) runs as follows. On a circular plate a radious vector is chosen at random and then either (a) the circumferential length of the smallest sector beginning at this vector and containing a prescribed number of colonies of bacteria, or (b) the total number of colonies on the plate, if they are fewer than the prescribed number, is used for estimating the density of bacteria per cc of the fluid population from which the sample on the plate has been taken. In case of (a) the choice of the counting direction should be independent of the actual distribution of colonies over the plate. Any colony on the initial vector should be excluded at the start of the counting procedure. It might instead be the last one counted. Two conditions are assumed to be satisfied: (1) The sampled volume is known except for a negligible technical error. (2) The sample has been spread out evenly on the plate, so that the number of colonies developed in equally large and non-overlapping sectors can be assumed to be observations on independent random variables with the same Poisson distribution. Instead of truncating the procedure one might use several plates, and also use different dilutions on these. A case when conditions (1) and (2) do not hold was considered in the third paper referred to above. A survey of statistical applications to bacteriology is given by [8]. For a discussion of errors in plate counts, see [14]; for related errors in blood cell counts, [4].

Some bacteriologists might prefer to have the plates divided into equally large sectors and to count colonies in whole sectors until the total count attains or exceeds a prescribed number S. "Equally large" should, in case of different dilutions, be read as "having the same expected number of colonies". The present paper will deal with this counting procedure in the non-truncated case, on the assumption that (1) and (2) hold.

The results presented might have applications outside bacteriology ([20], [1], [2], [11], [13], and chapter 2 of [6]). For this reason the following sections have been written in purely statistical terms. Although some of the formulas to be presented are believed to be new, most of them follow from known results by quite elementary arguments.

In Section 2 the sampling procedure is described and a distribution needed for the estimation is obtained. In Section 3 a very simple unbiased estimator of the Poisson parameter is obtained. An improved estimator based on the sufficient statistic is derived in Section 5. In Section 6 a randomized estimator is obtained by means of which a short efficiency study is made. In Section 7 the randomization device will be used for the construction of a simple estimator of the reciprocal of the Poisson parameter. Although the estimator considered in Section 3 can take the value zero, the probability of this event can be made arbitrarily small if S is chosen large enough. This will be shown in Section 4.

2. THE SAMPLING PROCEDURE AND A TRIVARIATE DISCRETE DISTRIBUTION ASSOCIATED WITH IT

To make the sampling procedure easy to understand we shall use different notations for random variables and observed values. We shall also let the word estimator denote a random variable and let the word estimate denote an observed value.

Let U_1, U_2, ... be a sequence of independent random variables, all having a Poisson distribution with mean λ. We make observations on the U's, say u_1, u_2, ..., respectively, until their sum, $\sum u_i$, attains or exceeds a given positive integer S. Obviously we must assume that

$$\lambda > 0. \tag{2.1}$$

In consonance with earlier practice ([31] and [10]) this sampling procedure will be called inverse Poisson sampling. Surveys of inverse sampling procedures have been made by [1] and [16]. The present procedure is mentioned by [19] (see page 3-40) as an example of a sequential procedure where the sufficient statistic lacks the completeness property.

The number of u's observed is an observation on a random variable n. The observed sum, $\sum u_i$, is an observation on a random variable which will be denoted by S_n. Further, if the last u_i is dropped, the remaining sum is an observed value of a random variable which will be denoted by S_{n-1}.

Let $f(x,y,z;\ \lambda,S)$ denote the probability that n takes the value x, S_{n-1} takes the value y, and S_n takes the value z, or

$$f(x,y,z;\ \lambda,S) = P(n = x,\ S_{n-1} = y \text{ and } S_n = z).$$

The possible values of x are the positive integers. The possible values of y are $0,1,\ldots,S-1$, and the possible values of z are $S,S+1,\ldots$. When u_1 exceeds $S-1$ we agree to say that S_{n-1} takes the value 0, so that in this case $x = 1$, $y = 0$, $z \geqq S$.

Hence

$$f(1,0,z;\ \lambda,S) = P(n = 1,\ S_{n-1} = 0 \text{ and } S_n = z)$$

$$= P(U_1 = z)$$

$$= \frac{e^{-\lambda}\lambda^z}{z!}\ .$$

For $x > 1$ we have, since for fixed x $\sum_{i=1}^{x-1} U_i$ and U_x are independent Poisson variables with means $(x-1)\lambda$ and λ, respectively,

$$f(x,y,z;\ \lambda,S) = P(n = x,\ S_{n-1} = y \text{ and } S_n = z)$$

$$= P\left(\sum_{i=1}^{x-1} U_i = y \text{ and } U_x = z - y\right)$$

$$= P\left(\sum_{i=1}^{x-1} U_i = y\right) \cdot P(U_x = z - y)$$

$$= \frac{e^{-(x-1)\lambda}[(x-1)\lambda]^y}{y!} \cdot \frac{e^{-\lambda}\lambda^{z-y}}{(z-y)!}$$

$$= \frac{e^{-x\lambda}(x-1)^y\lambda^z}{y!\ (z-y)!}\ ;\quad x = 2,3,\ldots;\quad y = 0,\ldots,S-1;\quad z = S,S=1,\ldots\ .$$

REMARK: Although, for given x, U_x is Poisson distributed, the random variable $S_n - S_{n-1}$ is not, since it cannot take the value zero.

An alternative way of deriving the joint frequency function for $x > 0$ is to observe that this function equals the probability that $\sum_{i=1}^{x} U_i$ takes the value z times the conditional probability that $\sum_{i=1}^{x-1} U_i$ takes the value y, given that $\sum_{i=1}^{x} U_i$ takes the value z. Hence

$$f(x,y,z;\lambda,S) = \frac{e^{-x\lambda}(x\lambda)^z}{z!} \cdot \binom{z}{y} \left(\frac{x-1}{x}\right)^y \left(\frac{1}{x}\right)^{z-y} . \tag{2.2}$$

The expression for the conditional probability follows from a well-known result found in [22].

Summing up, we have found that the joint distribution of the three discrete random variables n, S_{n-1} and S_n is

$f(x,y,z;\lambda,S) =$

$$P(n = x, S_{n-1} = y \text{ and } S_n = z) = \begin{cases} \dfrac{e^{-\lambda}\lambda^z}{z!}; & x = 1;\ y = 0;\ z = S,\ S+1, \ldots; \\ 0; & x = 1;\ y = 1, \ldots, S-1;\ z = S,\ S+1, \ldots; \\ \dfrac{e^{-x\lambda}(x-1)^y\lambda^z}{y!\,(z-y)!}; & x = 2,3,\ldots; \\ & y = 0,\ldots, S-1;\ z = S,\ S+1, \ldots . \end{cases} \tag{2.3}$$

The finiteness of the moments of this distribution follows from known results about cumulative sums and about the Poisson process. Thus all moments of n are finite, since the inverse sampling procedure is a special case of the sequential sampling procedure considered in [29]. Further, since all moments of the Poisson distribution are finite it follows from Theorem 2 of [35] that

$$E[S_n] = \lambda E[n]. \tag{2.4}$$

This is a well-known result of sequential analysis. (Cf. also [32], [33], [34], and [36].)

To prove that the higher moments of S_n and the mixed moments of n and S_n are finite we review some properties of the Poisson process. This review also includes some properties that we need later on.

A realization of a Poisson process is a sequence of events obtained by a certain chance mechanism and marked as points on a positive time axis. The following results about this chance mechanism are well-known:

(a) The origin and the successive events form intervals v_1, $v_2,\ldots$, which are observations on independent and exponentially distributed random variables $V_1, V_2, \ldots$, respectively, all with mean $1/\lambda$. Any event can serve as an origin, provided the choice is not influenced by the v_i that follow after that event.

(b) In non-overlapping intervals of lengths t_1 and t_2, the endpoints of which are chosen arbitrarily but independently of the realization, the number of events are observations on independent Poisson variables with means $t_1\lambda$ and $t_2\lambda$, respectively. In particular, the number of events in successive intervals $0<t\leq 1$, $1<t\leq 2$, etc., are observations on independent Poisson variables $U_1, U_2, \ldots$, all with mean λ. Any endpoint of such an interval can serve as an origin, provided the choice is not influenced by the observations after that endpoint.

(c) Given that, for fixed i, U_i takes the value u, the corresponding events are observations on independent variables all uniformly distributed over the interval $i-1 < t \leq i$.

(d) The distance w from i-1 to the j-th of the u events referred to under (c) is an observation on a random variable W which has the frequency function

$$g(w) = \frac{u!w^{j-1}(1-w)^{u-j}}{(j-1)!(u-j)!}, \quad 0 < w \leq 1;\ u = 1,2,\ldots;\ j = 1,\ldots,u, \tag{2.5}$$

the expected value $j/(u+1)$ and the second moment about zero $j(j+1)/[(u+1)(u+2)]$.

(e) The Poisson process can be simulated by simulating observations u_i on the U_i and then simulating u_i events in the respective interval using the property (c).

(For property (d) see, for instance, [38], pp. 13-14; for the other properties, see [6], ch. 2.)

Using (e) we assume that the S-th event has been simulated by means of the results of the inverse Poisson sampling procedure. By (a) and (e) the point denoting that event can be taken as a new origin. Let us assume that we simulate the events corresponding to the u-value that we would obtain if we supplemented the inverse Poisson sampling procedure by a new observation. Denoting the new origin by t_o, we let u' be the number of simulated events in the interval $t_o < t \leq t_o+1$. By (b) u' is an observation on a random variable, say U', which is Poisson-distributed with mean λ. Now $S_n - S$ cannot exceed U'. Hence

$$E\left[n^h(S_n-S)^k\right] < E\left[n^h U'^k\right], \quad h = 0,1,2,\ldots;\ k = 1,2,\ldots.$$

Now, by (a) the successive intervals between events beginning at t_o (i.e., $v_{S+1}, v_{S+2}, \ldots,$) are observations on $V_{S+1}, V_{S+2}, \ldots,$ respectively, which are independent of $\sum_{i=1}^{S} V_i$ and hence independent of n-1, the integral part of that sum. It follows that the right member of the above inequality equals the finite quantity

$$E[n^h] \cdot E[U'^k].$$

Hence all mixed moments of n and S_n are finite, and especially all moments of the latter.

As a by-product we get simple bounds on E[n]. For h = 0 and k = 1 we have that $E[S_n] < S+\lambda$. Further, since $S_n \geq S$, we have that $E[S_n] > S$. Hence, because $\lambda > 0$,

$$\frac{S}{\lambda} < E[n] < \frac{S}{\lambda}+1. \tag{2.6}$$

We end this section by giving a formula for E[n].

The cumulative distribution function of n, which is certainly known, is

$$F(x) = P(n \leq x) = P\left(\sum_{i=1}^{x} U_i \geqq S\right)$$

$$= \begin{cases} 0, & x = 0; \\ \sum_{z=S}^{\infty} \frac{e^{-x\lambda}(x\lambda)^z}{z!}, & x > 0. \end{cases} \tag{2.7}$$

REMARK: (2.7) can also be obtained from (2.3) by means of the well-known relation between the gamma and Poisson distributions.

Since n is an integer-valued random variable we have (see [9], p. 249) that

$$E[n] = \sum_{x=0}^{\infty} [1 - F(x)].$$

This yields

$$E[n] = 1 + \sum_{x=1}^{\infty} \sum_{k=0}^{S-1} \frac{e^{-x\lambda}(x\lambda)^k}{k!} . \tag{2.8}$$

The right member of (2.8) can be determined by means of a table of the cumulative distribution function of the Poisson distribution (see, e.g., [21]).

3. A SIMPLE UNBIASED ESTIMATOR OF THE POISSON MEAN

We shall prove that, for $S \geq 1$,

$$\lambda_1^* = \begin{cases} S_n, & n = 1 \\ \dfrac{S_{n-1}}{n-1}, & n > 1, \end{cases} \tag{3.1}$$

is an unbiased estimator of λ, and that, for $S \geq 2$,

$$V^*[\lambda_1^*] = \begin{cases} \lambda_1^*, & n = 1 \\ \dfrac{\lambda_1^*}{n-1}, & n > 1. \end{cases} \tag{3.2}$$

is an unbiased estimator of the variance of λ_1^*. Formulas (3.1) and (3.2) are believed to be new.

The estimate corresponding to λ_1^* is for $n > 1$, obtained by simply omitting the last observation and taking the mean of the remaining ones. In this case one can stop counting when a total count S is obtained and then record only the observed values of n and S_{n-1}.

For $S = 1$ we have, using (2.3) and observing that S_{n-1} is always zero in this case,

$$E[\lambda_1^*] = \sum_{z=S}^{\infty} z \, \frac{e^{-\lambda}\lambda^z}{z!} + 0 \cdot P(n > 1) = \lambda.$$

For $S > 1$ we have, writing y' for $y-1$ and z' for $z-1$,

$$E[\lambda_1^*] = \sum_{z=S}^{\infty} z \, \frac{e^{-\lambda}\lambda^z}{z!} + \sum_{x=2}^{\infty} \sum_{y=0}^{S-1} \sum_{z=S}^{\infty} \frac{y}{x-1} \, \frac{e^{-x\lambda}(x-1)^y \lambda^z}{y! \, (z-y)!}$$

$$= \lambda \sum_{x=1}^{\infty} \sum_{y'=0}^{S-1} \sum_{z'=S-1}^{\infty} f(x, y', z'; \lambda, S-1) = \lambda.$$

In a similar way we show that

$$\lambda_1^{2*} = \begin{cases} S_n(S_n - 1), & n = 1 \\ \dfrac{S_{n-1}(S_{n-1} - 1)}{(n-1)^2}, & n > 1, \end{cases} \tag{3.3}$$

is an unbiased estimator of λ^2. Hence

$$V^*[\lambda_1^*] = \lambda_1^{*2} - \lambda_1^{2*} \tag{3.4}$$

is an unbiased estimator of the variance λ_1^*. Since the right members of (3.2) and (3.4) are equal, (3.2) follows.

REMARK: (3.4) is an application of formula (1) of [12] which we reproduce here:

$$\text{est var } (f) = f^2 - \text{est } F^2; \tag{3.5}$$

here $F = E[f]$. (For other applications of (3.6) see, for instance, [10], [26], [27], and [30].)

4. ON THE PROBABILITY THAT λ_1^* TAKES THE VALUE ZERO

In the light of the assumption (2.1), that $\lambda > 0$, the fact that λ_1^* can take the value zero is an unpleasant property of this estimator. However, the probability of a zero-value can be made arbitrarily small. To show this we consider the function $g(\lambda)$, defined as follows:

$$g(\lambda) = P(\lambda_1^* = 0) = \sum_{x=2}^{\infty} \sum_{z=S}^{\infty} f(x,0,z;\ \lambda,S)$$

$$= \frac{e^{-\lambda}}{1-e^{-\lambda}} \sum_{z=S}^{\infty} \frac{e^{-\lambda}\lambda^z}{z!} = \frac{P(U = 0)P(U \geq S)}{P(U > 0)}, \tag{4.1}$$

where U denotes a Poisson variable with mean λ. Now we can write

$$g'(\lambda) = \frac{h(\lambda)}{[k(\lambda)]^2},$$

where

$$h(\lambda) = (1 - e^{-\lambda}) \frac{e^{-\lambda}\lambda^{S-1}}{(S-1)!} - \sum_{z=S}^{\infty} \frac{e^{-\lambda}\lambda^z}{z!}, \tag{4.2}$$

and

$$k(\lambda) = e^{-\lambda/2}(e^{\lambda} - 1).$$

Since (2.1) is assumed to hold true, the sign of $g''(\lambda)$ equals that of $h'(\lambda)$, when $h(\lambda) = 0$. Further, $h'(\lambda)$ can be simplified into

$$h'(\lambda) = \frac{(1 - e^{-\lambda})\ \lambda^{S-2}}{(S-1)!} (S-1-2\lambda).$$

By (2.1) $h'(\lambda)$ is negative if and only if

$$\lambda > \frac{S - 1}{2} . \tag{4.3}$$

Now the equation

$$h(\lambda) = 0 \tag{4.4}$$

is by (2.1) and (4.2) equivalent to

$$1 - e^{-\lambda} = \frac{\lambda}{S} \left(1 + \frac{\lambda}{S+1} + \frac{\lambda^2}{(S+1)(S+2)} + \ldots\right). \tag{4.5}$$

Hence

$$1 > 1 - e^{-\lambda} = \text{right member of (4.5)} > \frac{\lambda}{S} ,$$

so that

$$\lambda < S, \tag{4.6}$$

when λ satisfies (4.4). Further, since (2.1) implies that

$$e^{-\lambda} < \frac{1}{1 + \lambda} ,$$

we have, for $0 < \lambda < S$ and λ satisfying (4.5),

$$\frac{\lambda}{1+\lambda} = 1 - \frac{1}{1+\lambda} < 1 - e^{-\lambda} = \frac{\lambda}{S}\left(1 + \frac{\lambda}{S+1} + \frac{\lambda^2}{(S+1)(S+2)} + \ldots\right)$$

$$< \frac{\lambda}{S} \left(1 + \frac{\lambda}{S} + \frac{\lambda^2}{S^2} + \ldots\right) = \frac{\lambda}{S-\lambda} ,$$

so that

$$1 + \lambda > S - \lambda. \tag{4.7}$$

Now $h(\frac{S-1}{2}) > 0$ when $S > 1$. Since $h(S) < 0$ and $h(\lambda)$ is continuous, (4.4) has at least one solution for $S > 1$. But (4.7) and (4.3) are equivalent. Hence the solution is unique, and it yields the maximum of $P(\lambda_1^* = 0)$.

The numerical solution of (4.4) is easily obtained by trial and error with the help of the tables in [21] and the knowledge that the solution λ_o of (4.4) satisfies the inequalities (4.6) and (4.7). Some numerical results are given in Table 1. Substituting (4.4) into (4.1) we find

(1) that the maximum of $P(U = 0)P(U = S-1)$ with respect to λ provides an upper bound of Max $P(\lambda_1^* = 0)$;

(2) that this bound is obtained for $\lambda = (S-1)/2$; and

(3) that this bound approaches zero when S increases.

Hence Max $P(\lambda_1^* = 0)$ approaches zero when S increases. We see

Table 1. The Maximum of $P(\lambda_1^* = 0)$ and an Upper Bound of this Maximum

S	λ_o	Max. of $P(\lambda_1^* = 0)$	$\frac{S-1}{2}$	Max. of $P(U=0)P(U=S-1)$
2	0.85	0.160	0.5	0.184
3	1.5	0.055	1.0	0.068
4	2.1	0.0225	1.5	0.0280
5	2.7	0.0099	2.0	0.0122
10	5.4	0.000221	4.5	0.000258
15	7.9	0.0000058	7.0	0.0000065
20	10.4	0.000000157	9.5	0.000000174

from Table 1 that λ_1^*, for S = 10, will almost always be positive. As seen from the last column of the table the upper bound does not seriously overestimate Max $P(\lambda_1^* = 0)$.

5. AN IMPROVED ESTIMATOR OF THE POISSON MEAN

It follows from a general result (see [19], page 3-32) that the vector (n, S_n) is a sufficient statistic for λ. To get an estimator with a smaller variance than that of λ_1^* we shall determine the conditional expectation of λ_1^*, given that $(n, S_n) = (x,z)$. This gives an estimate which is an observation on an unbiased estimator λ_2^*. The idea of improving an estimator by determining its conditional expectation with respect to a sufficient statistic was presented in [3]. This idea is also implicitly used by [23] (see p. 281, lines 8 and 9 of that paper). The statistic need not be sufficient (see the note on p. 107 of [3], and also the theorem on p. 216 of [15]).

Using (2.2) and (2.3) we obtain

$$E[\lambda_1^* \mid n = x, S_n = z] = \frac{\sum_{y=0}^{S-1} \lambda_1^* f(x,y,z;\ \lambda, S)}{\sum_{y=0}^{S-1} f(x,y,z;\ \lambda, S)}$$

$$= \frac{z}{x} \cdot K(x,z), \tag{5.1}$$

where

$$K(x,z) = \begin{cases} 1, & x = 1 \\ \dfrac{\sum_{i=0}^{S-2} \binom{z-1}{i} \left(\frac{x-1}{x}\right)^i \left(\frac{1}{x}\right)^{z-1-i}}{\sum_{i=0}^{S-1} \binom{z}{i} \left(\frac{x-1}{x}\right)^i \left(\frac{i}{x}\right)^{z-i}}, & x > 1. \end{cases} \tag{5.2}$$

$K(x,z)$ can be easily computed by means of tables in [28] and [24], if we write, for $x > 1$,

$$K(x,z) = \frac{\sum_{i=z-S+1}^{z-1} \binom{z-1}{i} p^i (1-p)^{z-1-i}}{\sum_{i=z-S+1}^{z} \binom{z}{i} p^i (1-p)^{z-i}}, \tag{5.3}$$

where $p = 1/x$.

The first factor of (5.1) is the maximum-likelihood estimate of λ. Substituting n for x and S_n for z, we obtain the corresponding maximum-likelihood estimator

$$\hat{\lambda} = \frac{S_n}{n}. \tag{5.4}$$

This follows from the well-known fact that the likelihood function does not depend on the stopping rule used (cf., [37] p. 120). The improved estimator,

$$\lambda_2^* = \hat{\lambda} \cdot K(n,S_n), \tag{5.5}$$

can be considered as a bias-corrected ML-estimator.

Since, for $x > 1$,

$$0 < K(x,z) = 1 - \frac{\binom{z-1}{S-1} p^{z-S} (1-p)^{S-1}}{x \sum_{i=z-S+1}^{z} \binom{z}{i} p^i (1-p)^{z-i}} < 1, \quad z \geqq S, \tag{5.6}$$

it follows that $\hat{\lambda}$ is positively biased.

By the Blackwell-Rao device we also find that

$$\lambda_2^{2*} = \frac{S_n(S_n - 1)}{n^2} \cdot L(n,S_n), \tag{5.7}$$

where

$$L(x,z) = \begin{cases} 1, & x = 1 \\ \dfrac{\sum_{i=z-S+1}^{z-2} \binom{z-2}{i} p^i (1-p)^{z-2-i}}{\sum_{i=z-S+1}^{z} \binom{z}{i} p^i (1-p)^{z-i}}, & x > 1, \end{cases} \tag{5.8}$$

is an improved unbiased estimator of λ^2. Hence

$$V^*[\lambda_2^*] = \frac{S_n^2}{n^2} [K(n,S_n)]^2 - \frac{S_n(S_n - 1)}{n^2} L(n,S_n) \tag{5.9}$$

is an unbiased estimator of the variance of λ_2^*. It can be verified that

$$[k(x,z)]^2 > L(x,z) \tag{5.10}$$

Hence $V^*[\lambda_2^*]$ always takes a positive value. Except for formula (5.4) the formulas of this section are believed to be new.

6. A RANDOMIZED ESTIMATOR BY MEANS OF WHICH A LOWER BOUND ON ON THE EFFICIENCY OF λ_2^* IS OBTAINED

It follows from the properties of the Poisson process that we can construct a randomized estimator of λ by applying the simulation device mentioned in Section 2 to the last observation obtained by the inverse sampling procedure. We first review some well-known results. By the properties of the gamma distribution (cf. [7], Sect. 33.3, Ex. 3) we have the unbiased estimators

$$\lambda_3^* = \frac{S - 1}{\sum_{i=1}^{S} V_i}, \quad S > 1, \tag{6.1}$$

and

$$\lambda_3^{2*} = \frac{(S-1)(S-2)}{(\sum_{i=1}^{S} V_i)^2}, \quad S > 2, \tag{6.2}$$

of λ and λ^2, respectively, and also the unbiased estimator

$$V^*[\lambda_3^*] = \frac{S-1}{(\sum_{i=1}^{S} V_i)^2} = \frac{\lambda_3^{*2}}{S-1}, \quad S > 2, \tag{6.3}$$

of the variance of λ_3^*, which is

$$V[\lambda_3^*] = \frac{\lambda^2}{S-2}, \quad S > 2. \tag{6.4}$$

In these formulas $\sum_{i=1}^{S} V_i$ denotes the distance from the origin to the simulated S-th event of the Poisson process (cf. Section 2).

We now show that (5.5) and (5.7) can be obtained from (6.1) and (6.2), respectively, by the Blackwell-Rao device.

Let h(t) denote the conditional density function of $\sum_{i=1}^{S} V_i$, given that $(n, S_{n-1}, S_n) = (x,y,z)$. By (2.5)

$$h(t) = g(t-x+1),$$

where u and j in (2.5) are substituted by z-y and S-y, respectively. Hence the joint frequency-density function of (n, S_{n-1}, S_n) and $\sum_{i=1}^{S} V_i$ is, if $(x-1)^y$ is substituted by 1 for x = 1, y = 0,

$$f(x,y,z;\ \lambda,S)\cdot g(t-x+1) = \frac{e^{-x\lambda}\lambda^z (x-t)^{z-S}(x-1)^y (t-x+1)^{S-y-1}}{(z-S)!\ y!\ (S-y-1)!}$$

Summing this joint frequency function with respect to y, we obtain the joint frequency-density function of n, S_n and $\sum_{i=1}^{S} V_i$:

$$q(x,z,t) = \frac{e^{-x\lambda}\lambda^z t^{S-1}(x-t)^{z-S}}{(z-S)!\ (S-1)!}, \quad x = 1,2,\ldots;\ z = S, S+1, \ldots;\ x-1 < t \leq x. \tag{6.5}$$

It follows that the conditional frequency function of $\sum_{i=1}^{S} V_i$, given that $(n, S_n) = (x,z)$, is

$$a(t,x,z) = \frac{q(x,z,t)}{\int_{x-1}^{x} q(x,z,t)\,dt}. \tag{6.6}$$

By (6.5) and (6.6) and the well-known relation between the binomial and beta distributions, one can easily verify that the conditional expectation of λ_3^*, given that $(n, S_n) = (x,z)$, equals the right member of (5.1). The corresponding estimator (5.5) is thus verified. In a similar way we find that the conditional expectation of λ_3^{2*} equals the right member of (5.7), with x and z, instead of n and S_n, respectively.

Hence

$$V[\lambda_2^*] < V[\lambda_3^*] = \frac{\lambda^2}{S-2}. \tag{6.7}$$

We can now get a rough lower bound on the efficiency of λ_2^*. Let us assume that we stop the inverse Poisson sampling procedure after M observations, if not earlier, where M is a very large number. Then the lower bound of [39] on the variance of any unbiased estimator is applicable (cf. [19], page 2-36). Applying this bound to the Poisson distribution, we have that $\lambda/E[n]$ is a lower bound on any unbiased estimator of λ based on the inverse Poisson sampling procedure.

Hence the efficiency of λ_2^* is, by (2.6) and (6.7), at least

$$\frac{\frac{\lambda}{En}}{V[\lambda_2^*]} > \frac{S-2}{S+\lambda}. \tag{6.8}$$

The author is preparing a manuscript on the efficiencies of λ_1^* and λ_2^*. It appears that a numerical determination of the left member of (6.8) yields a much higher value than the one obtained from the right member, except when λ is small. A still higher efficiency is obtained by means of the bound of [5] (cf. also [18]).

7. UNBIASED ESTIMATORS OF $1/\lambda$

The randomization device also yields the following unbiased estimators of θ and θ^2, where $\theta = 1/\lambda$:

$$\theta_1^* = \frac{\sum_{i=1}^{S} V_i}{S} \tag{7.1}$$

and

$$\theta_1^{2*} = \frac{\left(\sum_{i=1}^{S} V_i\right)^2}{S(S+1)}. \tag{7.2}$$

Since

$$V[\theta_1^*] = \frac{\theta^2}{S}, \tag{7.3}$$

we also have the unbiased estimator

$$V^*[\theta_1^*] = \frac{\theta_1^{2*}}{S} = \frac{\theta_1^{*2}}{S+1} . \tag{7.4}$$

Formulas (7.1) - (7.4) are well-known and follow trivially from, e.g., the presentation of the Type III distribution given by [17] (see p. 55 therein).

Taking conditional expectation of θ_1^* and θ_1^{2*} with respect to n, S_{n-1}, and S_n, we obtain improved estimators of θ and θ^2, respectively. In this case we do not use the sufficient statistic alone. Since $\sum_{i=1}^{S} V_i$ can be written as n-1+W, we have by (2.5):

$$\theta_2^* = \frac{1}{S}\left(n - 1 + \frac{S - S_{n-1}}{S_n - S_{n-1} + 1}\right) \tag{7.5}$$

and

$$\theta_2^{2*} = \frac{1}{S(S+1)}\left((n-1)^2 + 2(n-1)\frac{S-S_{n-1}}{S_n-S_{n-1}+1} + \frac{(S-S_{n-1})(S-S_{n-1}+1)}{(S_n-S_{n-1}+1)(S_n-S_{n-1}+2)}\right), \tag{7.6}$$

respectively. We can improve these estimators by taking conditional expectation with respect to (n,S_n). The resulting expressions are quite complicated and are therefore omitted. Formulas (7.5) and (7.6) are believed to be new.

The main conclusion of the present paper is that the inverse sampling procedure yields fairly simple unbiased estimators of both the Poisson mean and its reciprocal.

REFERENCES

[1] Anscombe, F. J. 1953. Sequential estimation. J. R. Statist. Soc. B 15:1-21.

[2] Birnbaum, A. 1954. Statistical methods for Poisson processes and exponential populations. J. Amer. Statist. Assn. 49:254-6.

[3] Blackwell, D. 1947. Conditional expectation and unbiased sequential estimation. Ann. Math. Statist. 18:105-10.

[4] Chamberlain, A. C., and Turner, F. M. 1952. Errors and variations in white cell counts. Biometrics 8:55-65.

[5] Chapman, D. G., and Robbins, H. 1951. Minimum variance estimation without regularity assumptions. Ann. Math. Statist. 22:581-6.

[6] Cox, D. R., and Lewis, P. A. W. 1966. The Statistical Analysis of Series of Events. Methuen and Co., Ltd., London; Wiley and Sons, Inc., New York.

[7] Cramér, H. 1946. Mathematical Methods of Statistics. Princeton University Press, Princeton, N. J.

[8] Eisenhart, C., and Wilson, P. W. 1943. Statistical methods and control in bacteriology. Bacteriological Reviews 7:57-137.

[9] Feller, W. 1957. An Introduction to Probability Theory and Its Applications, 2nd ed. Wiley & Sons, Inc., New York.

[10] Finney, D. J. 1949. On a method of estimating frequencies. Biometrika 33:222-5.

[11] Girshick, M. A.; Rubin, H.; and Sitgreaves, R. 1955. Estimates of bounded relative error in particle counting. Ann. Math. Statist. 26:276-85.

[12] Glasser, G. J. 1962. On estimators for variances and covariances. Biometrika 49:259-62.

[13] Haight, F. A. 1967. Handbook of the Poisson Distribution. Wiley & Sons, Inc., New York, London, Sydney.

[14] Hedges, A. J. 1967. On the dilution errors involved in estimating bacterial numbers by the plating method. Biometrics 23:158-9.

[15] Hogg, R. V., and Craig, A. T. 1965. Introduction to Mathematical Statistics, 2nd ed. Macmillan, New York.

[16] Johnson, N. L. 1961. Sequential analysis: a survey. J. R. Statist. Soc. A 124:372-411.

[17] Kendall, M. G. 1943. The Advanced Theory of Statistics. Griffin & Co., Ltd., London.

[18] Kiefer, J. 1952. On minimum variance estimators. Ann. Math. Statist. 23:627-9.

[19] Lehmann, E. L. 1950. Notes on the Theory of Estimation. Associated Students Store. University of California, Berkeley, Calif.

[20] Malmquist, S. 1947. A statistical problem connected with the counting of radioactive particles. Ann. Math. Statist. 18:255-64.

[21] Molina, E. C. 1942. Poisson's Exponential Binomial Limit. Van Nostrand Co., Inc., New York, N. Y.

[22] Przyborowski, J., and Wilenski, H. 1940. Homogeneity of

results in testing samples from Poisson series with an application to testing clover seed for dodder. Biometrika 31:313-23.

[23] Rao, C. R. 1947. Minimum variance and the estimation of several parameters. Proceedings of the Cambridge Philosophical Society 43:280-3.

[24] Romig, H. G. 1953. 50-100 Binomial Tables. Wiley & Sons, Inc., New York, N. Y.

[25] Sandelius, M. 1950. An inverse sampling procedure for bacterial plate counts. Biometrics 6:291-2.

[26] ———. 1951. Inverse sampling applied to bacterial plate counts. I. Unrestricted and truncated sampling in the Poisson case. Kungl. Lantbrukshögskolans Annaler 18 86-94.

[27] ———. 1953. Inverse sampling applied to bacterial plate counts. II. Cases when technical errors cannot be neglected. Kungl. Lantbrukshogskolans Annaler 19:197-204.

[28] Statistical Engineering Laboratory 1949. Tables of the Binomial Probability Distribution. (National Bureau of Standards Applied Mathematics Series 6.) U. S. Government Printing Office, Washington, D. C.

[29] Stein, C. 1946. A note on cumulative sums. Ann. Math. Statist. 17:498-9.

[30] Sukhatme, P. V. 1954. Sampling Theory of Surveys with Applications. The Indian Society of Agricultural Statistics, New Delhi, India; The Iowa State College Press, Ames, Iowa.

[31] Tweedie, M. C. K. 1945. Inverse statistical variates. Nature 155:453.

[32] Wald, A. 1944. On cumulative sums of random variables. Ann. Math. Statist. 15:283-96.

[33] ———. 1945a. Sequential tests of statistical hypotheses. Ann. Math. Statist. 16:117-86.

[34] ———. 1945b. Sequential method of sampling for deciding between two courses of action. J. Amer. Statist. Assn. 40:277-306.

[35] ———. 1945. Some generalizations of the theory of cumulative sums of random variables. Ann. Math. Statist. 16:287-93.

[36] ———. 1947. Sequential Analysis. Wiley & Sons, Inc., New York, N. Y.

[37] Wetherill, G. B. 1966. Sequential Methods in Statistics. Methuen & Co., Ltd., London.

[38] Wilks, S. 1948. Order statistics. Bull. Amer. Math. Soc. 54:6-50.

[39] Wolfowitz, J. 1947. The efficiency of sequential estimates and Wald's equation for sequential processes. Ann. Math. Statist. 18:215-30.

SOME SIMPLE GRAPHICALLY ORIENTED STATISTICAL METHODS FOR DISCRETE DATA

JOHN J. GART

Biometry Branch
National Cancer Institute
Bethesda, Maryland

SUMMARY

In analyzing data it is often illuminating to have available arithmetically simple methods which have relevance to a graph of the data. This paper presents several such methods, some novel, which may provide useful insights in the analyses of certain kinds of counts involving various types of data.

1. INTRODUCTION

This paper is divided in three parts; the first is related to distributions (such as the Poisson distribution) of the non-negative integers; the second, to discrete counts of limited range such as the binomial distribution; and the last part is a special problem in the bioassay of viruses. Ord [23] and Dubey [10] have shown how certain simple algebraic properties of the various power series distributions give rise to certain useful graphs. In this paper we shall see how these graphs and modifications of them [14] are, in turn, related to estimators and statistical tests which are of full or nearly full efficiency. In some cases the test and estimators are well-known but apparently have not been previously related to these graphs. In the bioassay problem a modification of the Weibull graph is shown to lead to a fully efficient test [16].

2. POISSON-LIKE COUNTS

2.1 The Basic Graph

Consider the Poisson distribution

$$p(i) = \frac{\lambda^i e^{-\lambda}}{i!}, \quad i = 0,1,2,\ldots.$$

Ord [23] has pointed out the relationship

$$\frac{ip(i)}{p(i-1)} = \lambda, \; i = 1,2,3,\ldots, \tag{2.1}$$

which is easily shown to be unique for the Poisson. This suggests that if we are considering a random sample of non-negative counts, $\{x_j\}$, with the empirical frequencies f_i, $i = 0,1,\ldots,$ such that $\sum_{i=0}^{\infty} f_i = N$, then we consider the set of estimators,

$$\hat{\lambda}_i = \frac{if_i}{f_{i-1}}, \quad i = 1,2,3,\ldots$$

If $\hat{\lambda}_i$ is graphed against i, for i = 1,2,3,..., the points should cluster around the horizontal straight line $\bar{x} = \sum_{i=1}^{\infty} if_i/N$ when the data are adequately fitted by a Poisson distribution. It is noteworthy that the best estimator of λ, i.e., $\hat{\lambda} = \bar{x} = \sum_{i=1}^{\infty} if_i/N$, can be thought of as the weighted mean of the $\hat{\lambda}_i$'s,

$$\hat{\lambda} = \frac{\sum_{i=1}^{\infty} w_i \hat{\lambda}_i}{\sum_{i=1}^{\infty} w_i} ,$$

where the weights w_i are taken to be f_{i-1}.

2.2 The Truncated Poisson Distribution

It is immediately apparent that the basic relationship (2.1) extends itself to a truncated Poisson distribution,

$$p'(i) = \frac{\lambda^i/i!}{\sum_{j=a}^{b} \lambda^j/j!} \qquad i = a,\ldots,b,$$

where $0 \leq a < b \leq \infty$. Thus the graph of the individual estimators of $\hat{\lambda}_i$ are available for $i = a+1,\ldots,b$. In analogy with the full distribution, an estimator of λ can be taken as the weighted mean of the $\hat{\lambda}_i$'s,

$$\hat{\lambda} = \frac{\sum_{i=a+1}^{b} w_i \hat{\lambda}_i}{\sum_{i=a+1}^{b} w_i} .$$

If, as above, we take $w_i = f_{i-1}$, we find the estimator to be

$$\hat{\lambda} = \frac{\sum_{i=a+1}^{b} i f_i}{\sum_{i=a}^{b-1} f_i} = \frac{\sum_{i=a}^{b} i f_i - a f_a}{N - f_b} ,$$

or,

$$\hat{\lambda} = \frac{\bar{x} - a f_a/N}{1 - f_b/N} . \qquad (2.2)$$

If $a = 0$ then the estimator is

$$\hat{\lambda} = \frac{\bar{x}}{1 - f_b/N} , \qquad (2.3)$$

and if $b = \infty$ then the estimator is

$$\hat{\lambda} = \bar{x} - a f_a/N, \qquad (2.4)$$

which is an unbiased estimator of λ. The estimator in (2.3) was suggested by [20], and that in (2.4) by [27] for a = 1, and for general "a," by [32] and [19]. Subrahmaniam [32] also showed this estimator to be asymptotically nearly efficient for a wide variety of values of "a" and λ and gave an unbiased estimator of its variance.

The truncated estimators are useful when graphing a complete distribution since they can be used for estimates of λ after grouping the tails; (2.4) for the upper tail, and (2.3) for the lower tail.

2.3 Tests of the Poisson and Their Relationship to the Negative Binomial

The basic graph of the data described above can be used to construct tests of the Poisson model. It is reasonable to test whether any particular $\hat{\lambda}_i$ deviates significantly from $\hat{\lambda}$. For large samples it is easily found that $E(\hat{\lambda}_i-\hat{\lambda}) = 0$, and

$$V(\hat{\lambda}_i-\hat{\lambda}) = \frac{\lambda}{N}\left[\frac{\lambda}{p(i)} + \frac{\lambda}{p(i-1)} -1\right].$$ We may estimate the variance

by $$\hat{V}(\hat{\lambda}_i-\hat{\lambda}) = \frac{\hat{\lambda}}{N}\left[\frac{\hat{\lambda}}{\hat{p}(i)} + \frac{\hat{\lambda}}{\hat{p}(i-1)} - 1\right],$$

where $\hat{p}(j) = (\hat{\lambda}^j e^{-\hat{\lambda}})/j!$, $j = 0,1,2,\ldots$. A large sample normal deviate test can be based on

$$z = \frac{\hat{\lambda}_i-\hat{\lambda}}{\sqrt{\hat{V}(\hat{\lambda}_i-\hat{\lambda})}}. \tag{2.5}$$

Of course, if this test is applied to the most extreme deviation, the probability of finding significance is well above the prescribed significance level. Other similar tests are discussed somewhat more fully in [15]. All these tests may be useful in cases where the sample size is large and if the tests are used in the spirit of Cochran's [7] suggestion of breaking out a single degree of freedom when doing chi-square goodness-of-fit tests.

The test based on (2.5) does not involve a specific alternative to the Poisson distribution. A reasonable alternative is the negative binomial,

$$n(i) = \binom{r+i-1}{r-1}\frac{(m/r)^i}{(1+m/r)^{r+i}}, \quad i = 0,1,2,\ldots$$

for which $m > 0$, $r > 0$. When $r \to \infty$, the distribution becomes the Poisson distribution with mean m. The basic algebraic relationship corresponding to (2.1) is

$$\frac{i\, n(i)}{n(i-1)} = (r-1)\left(\frac{m/r}{1+m/r}\right) + \left(\frac{m/r}{1+m/r}\right) i, \; i = 1,2,\ldots. \tag{2.6}$$

That is, we have a linear relationship with positive slope, which, incidentally, must be less than one. This relationship has been pointed out in [23] (cf. [10]). We note, of course, that as $r \to \infty$, the right hand side of (2.6) goes to m, the mean of the corresponding Poisson distribution. The sample equivalent of (2.6) is the relationship

$$\hat{\lambda} = A + Bi, \; i = 1,2,\ldots; \tag{2.7}$$

that is, if the data are consistent with the negative binomial, the basic graph should be linear but with a positive slope. A test of the deviation of the data from the Poisson assumption in the direction of the negative binomial alternative can be formulated as H_o: $B = 0$, H_1: $1 > B > 0$. The weighted least squares estimator of B is

$$\hat{B} = \frac{\sum w_i (i-\bar{i}_w)(\hat{\lambda}_i - \bar{\lambda}_w)}{\sum w_i (i-\bar{i}_w)^2}, \tag{2.8}$$

where $\bar{i} = \sum w_i i \,/\, \sum w_i$ and $\bar{\lambda}_w = \sum w_i \hat{\lambda}_i \,/\, \sum w_i$.

If, as before, we chose $w_i = f_{i-1}$, we would find that (2.8) becomes

$$\hat{B} = 1 - \left(\frac{N}{N-1}\right)\frac{\bar{x}}{s^2}, \tag{2.9}$$

where s^2 is the usual unbiased estimator of the variance. Rejecting H_o when $\hat{B}$ is large and positive is seen to be equivalent to rejecting H_o when the Lexis ratio, $s^2/\bar{x}$, is large. But the Lexis ratio, or the variance ratio test, is the asymptotically locally optimal test of the Poisson distribution against the negative binomial alternative (see, e.g., [22]). It is, in fact, also optimal against any alternative to the Poisson found by treating λ as itself a random variable.

Thus we have seen that a "best" test of Poisson distribution can be related to a very simple and informative graph of the data and need not be considered in abstraction.

2.4 Examples of the Basic Graph

We reproduce in Table 1 the classical data on radioactive counts of [29]. The $\hat{\lambda}_i$'s are also shown there and are graphed in Figure 1. The truncated estimator (2.4) is used for $i \geq 11$. The $\hat{\lambda}_i$'s are seen to cluster around the horizontal line, $x = 3.870$. No trend is observed in them, although $\hat{\lambda}_8 = 2.590$ seems to fall appreciably below the horizontal line and $\hat{\lambda}_9$ seems to fall appreciably above the horizontal line. If we apply the normal deviate test of (2.5) we find for $\hat{\lambda}_8$ that $z = -2.24$ and for $\hat{\lambda}_9$ that $z = +2.01$, both of which appear to be significant on an individual basis. However, since $\hat{\lambda}_8$ and $\hat{\lambda}_9$ are negatively correlated and were chosen as the most extreme from a set of 11 possible tests, these significance levels must be viewed with caution. On the other hand Berkson [3] has pointed out that these data have a Lexis ratio $s^2/\bar{x} = 0.955$, which is a borderline evidence ($P = 0.049$ for a one-tailed test) of an undispersion in a direction opposite to the negative binomial, but is consistent with the model of the Type 1 counter discussed in [9] (pp. 31-32 and 40-42).

Table 1. Distribution of the Number of α Particles Radiated from a Disc in 7.5 Seconds (Source: [29])

i	f_i	$\hat{\lambda}_i$	
0	57		
1	203	3.561	
2	383	3.773	
3	525	4.112	
4	532	4.053	
5	408	3.835	
6	273	4.015	
7	139	3.564	
8	45	2.590	
9	27	5.400	
10	10	3.704	
11	4	4.400	
12	2	6.000	4.125
≥ 13	0	0.000	
	2608	$\hat{\lambda}. = 3.870$	

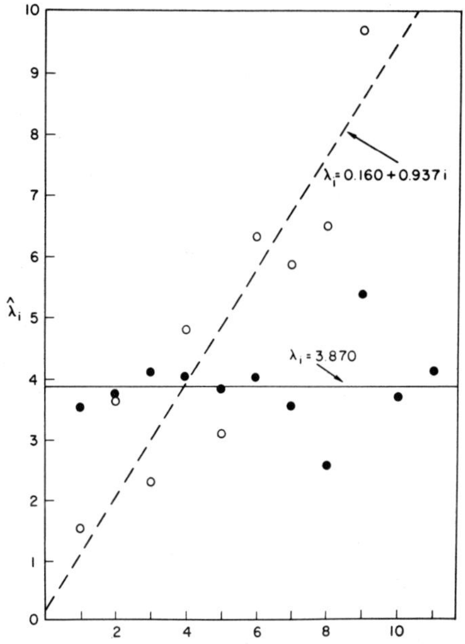

Figure 1
Graph of α Particle Counts
(o,——)
and
Bacteria Counts
(o,---)

Table 2. Distribution of Soil Bacteria (Source: [17])

i	f_i	$\hat{\lambda}_i = \frac{i\,f_i}{f_{i-1}}$	
0	11		
1	17	1.545	
2	31	3.647	
3	24	2.323	
4	29	4.833	
5	18	3.103	
6	19	6.333	
7	16	5.895	
8	13	6.500	
9	17	11.769	
10	6	3.529	9.710
11	8	14.667	
12	31		

Our second example, in Table 2, is the data of [17] on the distribution of soil micro-organisms (see also [4]). The points obviously show a trend; the least squares estimator of B is 0.937, or very close to the upper bound of 1 for B. This reflects the fact that the Lexis Ratio is large, $s^2/\bar{x} = 15.94$; that is, the homogenity [7] chi-square is almost sixteen times its theoretical expectation of 239.

2.5 Estimation in the Complete and Truncated Negative Binomial Distributions

The relationships (2.6) through (2.9) may also be used to obtain estimators of the complete negative binomial distribution.

The estimator of the intercept (2.7) is

$$\hat{A} = \bar{\lambda}_w - \hat{B}\bar{i}_w .$$

If we choose $w_i = f_{i-1}$ as before and equate $\hat{A}$ and $\hat{B}$ to their corresponding theoretical values under the negative binomial model (2.6), we find the estimators

$$\hat{m} = \bar{x},$$

$$\hat{r} = \frac{N\bar{x}^2}{(N-1)s^2 - N\bar{x}},$$

which are, of course, the method of moments estimators. Fisher [12] has shown in what instances these estimators may have high efficiency.

The above method can also be employed to derive a more useful set of estimators for the truncated negative binomial distribution. We consider the left truncated case explicitly:

$$n'(i) = C \binom{r+i-1}{r-1} \frac{(m/r)^i}{(1+m/r)^{r+i}}, \quad i = a, a+1, \ldots,$$

where

$$C = \frac{1}{\sum_{i=a}^{\infty} n(i)} .$$

Of course, the linear relationship of (2.6), holds for this distribution, and regression estimators when $w_i = f_{i-1}$, $i = a+1, a+2, \ldots$, yield

$$\hat{B}' = 1 - \frac{a(a-1)f_a + (N-af_a)\bar{x}}{(N-1)s^2}$$

$\hat{A}' = \bar{x} - afa/N - \hat{B}'(\bar{x}+1)$.

Equating these estimators to their theoretical values in (2.6) and solving, we find

$$\hat{m}' = x - \left(\frac{N-1}{N}\right) \frac{af_a s^2}{[a(a-1)f_a + (N - af_a)\bar{x}]}, \qquad (2.10)$$

$$\hat{r}' = \frac{(N-af_a)\bar{x}^2 + a(a-1)f_a\bar{x} - [(N-1)/N]af_a s^2}{(N-1)s^2 - a(a-1)f_a - (N-af_a)\bar{x}} \qquad (2.11)$$

The efficiency of these estimators has not been investigated in general. However, when a = 1, $\hat{m}'$ and $\hat{r}'$ are practically equivalent (except for the usual N-1 corrections) to those considered by Brass [5]. For $m \leq r$ he found that the efficiency was quite high (> 90%). However, when $m > r$, its efficiency can be quite low. For the a = 1 case, [30] suggested a set of estimators based on the first two moments, which require iterative calculations. [5] Shows that when $m > r$ the estimators of (2.10) and (2.11) are relatively much more efficient than the moment estimators, and when $r > m$ they are only slightly less efficient than the moment estimators. Thus these estimators seem to be preferable to the moment estimators both on the grounds of simplicity and efficiency.

It is possible to use the same kind of regression relationship to find estimators of the negative binomial parameters in the right-truncated and doubly truncated cases. The explicit form of such estimators and their efficiencies will require further research.

Example of Truncated Binomial: Sampford considered the following data of C. E. Ford on chromosome breaks in irradiated tissue, where the zero class is truncated: $f_1 = 11$, $f_2 = 6$, $f_3 = 4$, $f_6 = 1$, $f_8 = 2$, $f_9 = f_{11} = f_{13} = 1$, and all other observed frequencies are zero. Using the method of moments and iterative techniques, after 15 iterations Sampford found an estimate of m to be 2.065 and of r, 0.633. Application of (2.10) and (2.11) yields the estimates, $\hat{m}' = 1.971$ and $\hat{r}' = 0.604$. The latter estimators may be used as initial estimators in the iterative likelihood scheme of Sampford to yield the estimators for m of 1.840, and for r, 0.493.

2.6 The Log-series and the Truncated Log-series

The logarithmic series distribution of Fisher is defined by

$$s(i) = \frac{\alpha\,\theta^i}{i}, \quad i = 1,2,3,\ldots,$$

where $\alpha = k/[\ln(1-\theta)]$ and $0 < \theta < 1$. A slight modification of the Poisson plot is appropriate here. It is easily found that

$$\frac{is(i)}{(i-1)s(i-1)} = \theta, \text{ for } i = 2,3,\ldots,$$

a relationship closely related to that pointed out by Ord. The individual estimators of the θ are,

$$\hat{\theta}_i = \frac{i\ f_i}{(i-1)\ f_{i-1}}, \quad i = 2,3,\ldots;$$

that is, the $\hat{\theta}_i$'s should cluster around a horizontal straight line when the data fit the log-series model. A simple estimator of θ is found by taking weighted means of the $\hat{\theta}_i$'s, $\hat{\theta} = \sum w_i\hat{\theta}_i/\sum w_i$; if $w_i = (i-1)f_{i-1}$, for $i = 2,3,\ldots$, then we find

$$\hat{\theta} = 1 - \frac{f_1}{N\bar{x}},$$

which is the "first cell-first moment" estimator suggested by Patil [24]. Patil showed this estimator to be the best among the three simple estimators he investigated, with asymptotic efficiencies generally above 80% and becoming fully efficient as $\theta \to 0$. However, since $\bar{x}$ is a sufficient statistic for θ and the solutions to the likelihood equation have been tabled in [26], this simple estimator is not of much practical use.

The plot can be used to assess fit of the log-series model and to construct simple estimators for the truncated case. If the log-series is truncated so as to range from a to b, we find a simple estimator to be

$$\hat{\theta} = \frac{\bar{x} - af_a/N}{\bar{x} - bf_b/N}.$$

If $b \to \infty$ this becomes

$$\hat{\theta} = 1 - \frac{af_a}{N\bar{x}}.$$

If $a = 1$, this becomes

$$\hat{\theta} = \frac{\bar{x} - f_1/N}{\bar{x} - bf_b/N} .$$

Some tables for the maximum likelihood estimator of the latter case have been given by [26].

The efficiencies of these various truncated cases are still to be investigated.

3. BINOMIAL-LIKE COUNTS

3.1 The Basic Graph

We consider first the binomial distribution,

$$b(i) = \binom{m}{i} p^i q^{m-i}, \quad i = 0,1,\ldots,m,$$

where $p + q = 1$. Ord [23] (see also [10]) has noted the relationship

$$\frac{ib(i)}{b(i-1)} = (m+1)\,\frac{p}{q} - \frac{p}{q}\,i, \quad i = 1,2,\ldots$$

Gart [14] suggested a simpler and more useful relationship

$$p = \frac{ib(i)}{ib(i) + (m-i+1)b(i-1)}, \quad i = 1,2,\ldots,m. \qquad (3.1)$$

This suggests that if we are analyzing a random sample of counts $\{x_i\}$ with empirical frequencies f_i, $i = 0, 1,\ldots,m$, such that $\sum f_i = N$, then we consider the set of estimators

$$\hat{p}_i = \frac{if_i}{(m-i+1)f_{i-1}+if_i}, \quad i = 1,2,\ldots,m. \qquad (3.2)$$

A logical estimator of p is the weighted mean $\hat{p} = \sum w_i \hat{p}_i / (\sum w_i)$. If we take $w_i = (m-i+1)f_{i-1}+f_i$, then $\hat{p} = \left(\sum_{i=1}^{m} if_i\right)/Nm = \bar{x}/m$, which is, of course, the best estimator. As with the Poisson, if we graph the $\hat{p}_i$'s against i they should cluster around the horizontal straight line with ordinate $\hat{p}$.

3.2 The Truncated Binomial

The relationship (3.1) obviously holds for binomial distributions which are truncated in any way. If the range is restricted so that $a \le x_i \le b$, then linear weighted mean of the $\hat{p}_i$'s, using the above weights, is

$$\hat{p} = \frac{\sum_{i=a+1}^{b} w_i \hat{p}_i}{\sum w_i} = \frac{\bar{x} - (afa)/N}{m - [af_a + (m-b)f_b]/N} . \tag{3.3}$$

When a = 0, (3.3) becomes

$$\hat{p} = \frac{\bar{x}}{m - [(m-b)\ f_b]/N} , \tag{3.4}$$

and when b = m, (3.3) becomes

$$\hat{p} = \frac{\bar{x} - af_a/N}{m - af_a/N} . \tag{3.5}$$

Two special cases of the truncated binomial are of particular relevance in human genetics. The doubly truncated case a = 1, b = m-1, for which (3.3) becomes (see [14],

$$\hat{p} = \frac{\bar{x} - f_1/N}{m - (f_1 + f_{m-1})/N} ,$$

and the singly truncated case, a = 1, b = m, the so-called "complete ascertainment case", for which we find from (3.3):

$$\hat{p} = \frac{\bar{x} - f_1/N}{m - f_1/N} . \tag{3.6}$$

The latter estimator was first suggested in [19] using a different rationale in connection with a problem in diagnostic tests. Gart [14] (see also [18]) investigated its asymptotic relative efficiency and found it to be at least 95% efficient for a broad spectrum of values of m and p.

The complete ascertainment case is of such practical significance that [25] has produced a rather complete table from which the maximum likelihood estimator of p may be directly read. The question arises whether the simple estimator given in (3.6) is of any practical value when an asymptotically efficient alternative is so easily available. A recent study ([33]) has shown that the simple estimator is less biased than the maximum likelihood estimator for finite sample sizes and, in fact, that it is sometimes more efficient than the maximum likelihood estimator in small samples when efficiency is measured by the ratio of the mean square errors.

A generalized form of the simple estimator is of particular usefulness when combining data for varying m's (i.e., family sizes).

3.3 Tests of the Binomial and Their Relationship to the Basic Graph

As with the Poisson graph it is reasonable to consider the deviations $\hat{p}_i - \hat{p}$, in constructing tests of fit of a binomial. We find for large samples that

$$V(\hat{p}_i - \hat{p}) = \frac{pq}{N}\left[\frac{pq}{b(i)} + \frac{pq}{b(i-1)} - 1\right],\ i = 1,2,\ldots,m, \tag{3.7}$$

We may estimate this variance in the usual way and use these results to construct approximate normal deviate tests, as was done with the Poisson.

If we wish to consider an alternative to the binomial, a tractable form is found by assuming that p has a beta-distribution, integrating the p out and obtaining the compound binomial or beta-binomial distribution:

$$c(i) = \binom{m}{i}\frac{B(\alpha+i,\beta+m-i)}{B(\alpha,\beta)},\ i = 0,1,\ldots,n, \tag{3.8}$$

where $\beta > 0$, $\alpha > 0$. When $\alpha \to \infty$ and $\beta \to \infty$, such that $\alpha/(\alpha+\beta) = p$, then $c(i) \to b(i)$ with parameter p. If we apply the basic graph of (3.1) to c(i), we find

$$\frac{ic(i)}{ic(i) + (m-i+1)c(i)} = \frac{\alpha-1}{\alpha+\beta+m-1} + \frac{i}{\alpha+\beta+m-1}, \tag{3.9}$$

for $i = 1,2,\ldots,m$. That is, we have a linear relationship with positive slope which must be less than $(m-1)^{-1}$. We note that, in the limiting case noted above, the right-hand side of (3.9) approaches p in correspondence with (3.1). The simple equivalent of (3.8) is

$$\hat{p}_i = A+Bi,\ i = 1,2,3,\ldots; \tag{3.10}$$

that is, if the data deviate from the binomial in the direction of the beta-binomial, the basic graph should be linear with positive slope. The hypothesis can be formulated in terms of H_o: $B = 0$, H_1: $0<B<(m-1)^{-1}$. The weighted least squares estimator of B with weights chosen above is

$$\hat{B} = \frac{X^2-N}{(m-2)X^2+N},$$

where X^2 is the homogeneity chi-square ([7]), for the binomial

$$X^2 = \frac{\sum(x_i-\bar{x})^2}{m\hat{p}\hat{q}}.$$

It can be easily shown that, for $m \geq 3$ and $\hat{B} < (m-2)^{-1}$, X^2 is a monotone increasing function of $\hat{B}$, and thus a test based on $\hat{B}$ is equivalent to the homogeneity chi-square test. Of course, the latter is the asymptotically locally optimal test for the binomial against the general alternative of p being itself a random variable (e.g., [28]). Thus, we see once again that a "best" test of the binomial can be related to a simple and informative graph.

3.4 Examples of the Basic Graph

In Table 3 we reproduce the classic dice data of Weldon (e.g., [13], p. 64). Here $m = 12$, and the theoretical $p = 1/3$. The $\hat{p}_i$'s are entered for each $i = 1,2,\ldots,m$; because of the small numbers beyond $i = 10$, the truncated estimator of (3.5) is given for $a = 10$. The $\hat{p}_i$'s are graphed in Figure 2, and the points are seen to cluster about the horizontal line $\hat{p} = 0.3377$. The largest deviation is $\hat{p}_9 - \hat{p} = 0.0319$; however the test described in Section 3.3 yields a normal deviate for this of only 1.26.

The data of Geissler on the distribution of sex distribution in 6115 families of size twelve is reproduced in Table 4, together with the corresponding $\hat{p}_i$'s. The graph in Figure 2 is seen to have an approximately linear trend. For these data, $\hat{B} = 0.0123$, which corresponds to a homogeneity chi-square of 7058 with 6114

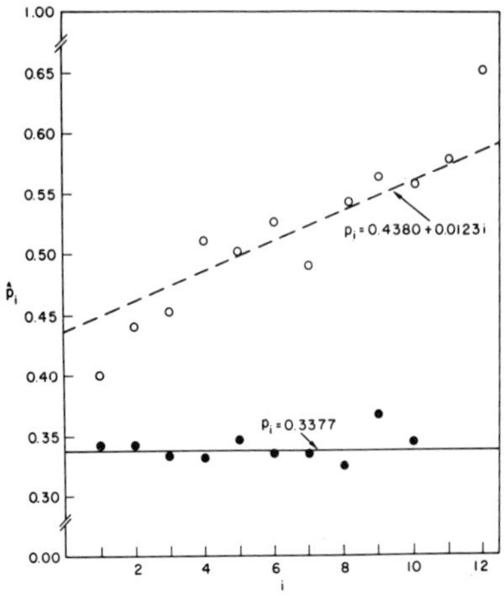

Figure 2
Graph of Dice Data (o,——)
and
Sex Ratio Data (o,---)

Table 3. Weldon's Dice Data
(Source: [13])

i	f_i	$p_i = \frac{if_i}{if_i + (13-i)\ f_{i-1}}$	
0	185		
1	1149	0.3411	
2	3265	0.3407	
3	5475	0.3347	
4	6114	0.3417	
5	5194	0.3368	
6	3067	0.3360	
7	1331	0.3361	
8	403	0.3263	
9	105	0.3696	
10	14	0.3077	
11	4	0.6111	0.3465
12	0	0.0000	
	N = 26,306	$\hat{p}$ = 0.3377	

Table 4. Geissler's Data on the Sex Distribution in Families
(Source: [11])

No of Boys i	No of Families f_i	$\hat{p}_i = \frac{if_i}{if_i + (13-1)f_{i-1}}$
0	3	
1	24	0.4000
2	104	0.4407
3	286	0.4521
4	670	0.5101
5	1033	0.5002
6	1343	0.5270
7	1112	0.4914
8	829	0.5440
9	478	0.5647
10	181	0.5580
11	45	0.5776
12	7	0.6512
	N = 6115	$\hat{p}$ = 0.5192

degrees of freedom and is highly significant. Edwards [11] has fitted a beta-binomial to these data.

3.5 Estimation for the Beta-binomial and the Truncated Beta-binomial

As with the negative binomial the basic graph can be used to estimate the parameters of the beta-binomial distribution. The linear regression of $\hat{p}_i$ on i, with the usual weights, leads to the moment estimators of α and β suggested by [31] and used by [11] in fitting Geissler's data. Corresponding estimators for the truncated cases may be easily derived.

As with the negative binomial, these estimators are not necessarily very efficient. The $\hat{p}_i$'s are, of course, correlated random variables; the appropriate regression estimators should take account of the correlations in the $\hat{p}_i$'s in fitting the straight line. The covariance matrix of the $\hat{p}_i$'s has been derived; and the question of more efficient estimation for this case, as well as the negative binomial case, will be elucidated in a later paper.

4. HOST VARIABILITY AND THE SINGLE PARTICLE CURVE

4.1 The Basic Model and Derivation of Test Statistic

The concept of the single particle (or one-hit curve) is a useful one in interpreting the possible mechanism of action of a virus in a living host. The experimental situation involves a series of dilutions, d_i, $i = 1,2,\ldots,k$, of a virus solution at each of which n hosts (mice, chick embryos, etc.) are exposed to a given volume of the material. The number of virus particles is assumed to follow a Poisson distribution with mean λd_i. If a single virus particle is invariably enough to cause death (or some other response) in the host by some prescribed time, then the probability of survival of the host at dose d_i is the zero term of the Poisson distribution:

$$P_i = \exp(-\lambda d_i), \; i = 1,2,\ldots,$$

the so-called single particle curve (see for instance [6] or [8]). It may be, however, that because of the natural defenses of the host, each virus has only a probability p of leading to the host's death. Then it is easily shown that

$$P_i = \exp(-\lambda p d_i),\ i = 1,2,\dots,k, \tag{4.1}$$

a curve of the same form as above.

If, in addition, the host population varies randomly in their susceptibility to the virus so that p has a distribution function F(p), then we find,

$$P_i = \int_0^1 \exp(-\lambda p d_i)\,dF(p),\ i = 1,2,\dots,k. \tag{4.2}$$

Armitage [1] assumed the F(p) to be a type III distribution or the truncated exponential and derived the asymptotically locally optimal test of the model of (4.1) against the alternative of (4.2). This test is very difficult to compute and is not associated with any particular graph.

Gart and Weiss [16], starting with a more general formulation of F(p) given by [2], derived a fully efficient, easily computable and graphically oriented test. If we expand the exponential in (4.2) about its mean, $E(p) - \bar{p}$ and integrate, we obtain

$$P_i = e^{-\lambda p d_i}\left[1+(\lambda d_i)^2\,\frac{\mu_2}{2!} - (\lambda d_i)^3\,\frac{\mu_3}{3!} + \dots\right],$$

where $\mu_j = E(p-\bar{p})^j$ for all j. If we neglect all higher moments beyond the second, we have

$$P_i \sim e^{-\gamma\, d_i}\ 1+\frac{V}{2}(\gamma d_i)^2, \tag{4.3}$$

where $\gamma = \lambda\bar{p}$ and $V = \mu_2/\bar{p}^2$. We are concerned with the test of whether the square of the coefficient of variation of p is zero: $V = 0$, against the alternative $V > 0$. Taking natural logarithms of (4.3), we have

$$\ln P_i \sim -\gamma d_i + \ln[1+\frac{V}{2}(\gamma d_i)^2].$$

Neglecting terms of order $O(V^{-2})$, we have

$$\ln P_i \sim -\gamma d_i[1-\frac{V}{2}(\gamma d_i)].$$

Taking logarithms and expanding in a series again, we find,

$$\ln(-\ln P_i) \sim \ln\gamma + \ln d_i - \frac{V}{2}(\gamma d_i).$$

Transposing, we may write this as,

$$\ln(-\ln P_i) - \ln d_i \sim \ln\gamma - \frac{V}{2}(\gamma d_i). \tag{4.4}$$

We notice that the left hand side of (4.4) can be estimated from the data and the prescribed constants, while the right hand side is a linear function of dilutions, d_i, with negative slope. Under the model of (4.1), V = 0, and the linear graph is horizontal. This suggests a simple test of the hypothesis of host variability; regress $Z_i = \ln(-\ln\hat{P}_i) - \ln d_i$ on d_i, and test whether the slope is significantly negative. The regression is a weighted one, with weights given by

$$w_i = \frac{1}{V[\ln(-\ln\hat{P}_i)]} = \frac{n_i P_i \ln^2 P_i}{1-P_i} .$$

Further details and examples of the test are given in [16].

The interesting fact is that, although this test was constructed on the basis of a series of mathematical approximations and using heuristic reasoning, its asymptotic local efficiency matches that of the more elaborate test constructed by [1] using the sophisticated general theory of [21]. More important, its computation is relatively easy and it can be interpreted in light of a simple graph, which was termed the modified Weibull Plot in [16].

5. CONCLUSION

5.1 Final Remarks

The theoretical structure of the science of statistics owes much to those who have propounded unifying principles and constructed general methods of analyzing data. Fisher, in statistical estimation, and Neyman and Pearson, in the statistical testing, have provided general tools that will be useful in almost any problem. In many situations these methods will greatly aid a research worker in the interpretation of his data. However on some occasions the general theory leads to estimators and tests of great algebraic and arithmetic complexity. Although some research workers may be impressed by 'Type A Critical Regions,' 'iterative likelihood solutions,' and 'locally asymptotically most powerful tests,' they may not be equally enlightened.

It has been the intention of this paper to indicate how some of the 'best' tests and estimators as defined by these general theories can be related to some very simple graphs. We have also indicated, in one instance, that a test related to a simple graphical method is as efficient as the 'optimal' test based on a general theory. Hopefully, in other situations statistical analyses and simple graphs may be similarly wedded to produce a better understanding of data.

REFERENCES

[1] Armitage, P. 1959. Host variability in dilution experiments. Biometrics 15:1-9.

[2] ———., and Spicer, C. C. 1956. The detection of variation in host susceptibility in dilution counting experiments. The Journal of Hygiene (Cambridge) 54:401-14.

[3] Berkson, J. 1966. Examination of randomness of α-particle emissions. Research Papers in Statistics, Festschrift for J. Neyman. (F. N. David, ed.) New York, Wiley, 37-54.

[4] Bliss, C. I. 1953. Fitting the negative binomial to biological data. Biometrics 19:176-96.

[5] Brass, W. 1958. Simplified methods of fitting the truncated negative binomial distribution. Biometrika 45:59-68.

[6] Cochran, W. G. 1950. Estimation of bacterial densities by means of the 'most probable number'. Biometrics 6:105-16.

[7] ———. 1954. Some methods of strengthening the common chi-square tests. Biometrics 10:417-51.

[8] Cornfield, J. 1954. Measurement and comparison of toxicities: the quantal response, Statistics and Mathematics in Biology, (O. Kempthorne et al., eds.) Iowa State Press, 327-44.

[9] Cox, D. R. 1962. Renewal Theory, London, Hafner.

[10] Dubey, S. D. 1966. Graphical tests for discrete distributions. The American Statistician 20:23-24.

[11] Edwards, A. W. F. 1958. An analysis of Geissler's data on the human sex ratio. Ann. Hum. Genet., London 23:6-15.

[12] Fisher, R. A. 1941. The negative binomial distribution, Ann. Eugenics 11:182-7.

[13] ———. 1958. Statistical Methods for Research Workers. New York, Hafner.

[14] Gart, J. J. 1968. A simple, nearly efficient alternative to the simple sib method in the complete ascertainment case. Ann. Hum. Genet., London 31:283-91.

[15] ———. 1969. Graphically oriented tests of the Poisson distribution. Presented at the 37th Session of the International Statistical Institute, London.

[16] ————., and Weiss, G. H. 1967. Graphically oriented tests for host variability in dilution experiments. Biometrics 23:269-84.

[17] Jones, P. C. T.; Mollison, J. E.; and Quenouille, M. H. 1948. A techinque for the quantitative estimation of soil micro-organisms. J. Gen. Microbiology 2:54-69.

[18] Li, C. C. and Mantel, N. 1968. A simple method of estimating the segregation ratio under complete ascertainment. Amer. J. Hum. Genet. 20:61-81.

[19] Mantel, N. 1951. Evaluation of a class of diagnostic tests. Biometrics 3:240-46.

[20] Moore, P. G. 1952. The estimation of the Poisson parameter from a truncated distribution. Biometrika 39:247-51.

[21] Neyman, J. 1959. Optimal tests of composite statistical hypotheses, Probability and Statistics, the Harald Cramer Volume (U. Grenander, ed.) New York: John Wiley and Sons.

[22] ————., and Scott, E. 1965. On the use of $C(\alpha)$ optimal tests of composite hypothesis. Proc. Int. Stat. Inst. 35, paper no. 118.

[23] Ord, J. K. 1967. Graphical methods for a class of discrete distributions. J. Roy. Stat. Soc. A, 130:232-38.

[24] Patil, G. P. 1962a. Some methods of estimation for the logarithmic series distribution. Biometrics 18:68-75.

[25] ————. 1962b. Maximum likelihood estimation for generalized power series distributions and its application to a truncated binomial distribution. Biometrika 49:227-37.

[26] ————., and Wani, J. K. 1965. Maximum likelihood estimation for the complete and truncated logarithmic series distributions. Sankhya 27:281-292.

[27] Plackett, R. L. 1953. The truncated Poisson distribution. Biometrics 9:485-8.

[28] Potthoff, R. F., and Whittinghill, M. 1966. Testing for homogeneity I. The binomial and multinomial distributions. Biometrika 53:167-82.

[29] Rutherford, E., and Geiger, H. 1910. Phil. Mag. Ser. 6, 20:698.

[30] Sampford, M. R. 1955. The truncated negative binomial distribution. Biometrika 42:58-69.

[31] Skellam, J. G. 1948. A probability distribution derived from the binomial distribution by regarding the probability of a success as variable between sets of trials. J. Roy. Stat. Soc. B, 10:257-61.

[32] Subrahmaniam, K. 1965. A note on estimation in the truncated Poisson. Biometrika 52:279-82.

[33] Thomas, D. G., and Gart, J. J. 1968. The small sample performance of some estimators of the truncated binomial distribution. Presented ASA National Meeting, Pittsburgh, Pa., Aug. 1968. (submitted for publication in J. Amer. Stat. Assoc.)

GRAPHICAL METHODS FOR THE DETERMINATION OF TYPE AND PARAMETERS OF SOME DISCRETE DISTRIBUTIONS

H. GRIMM

Institute for Microbiology and Experimental Therapy
Jena, Germany

SUMMARY

The types of some discrete distributions are readily discerned by plotting data on Poisson cumulative probability paper. In this paper this subject is discussed in depth, showing how, among other means, transparent stencils may be employed to quickly recognize typical curves and estimate their parameters.

1. INTRODUCTION

Numerical methods are often used in practical applications to find the distribution which characterizes random counts, but such methods are invariably laborious and time consuming. Neither the index of dispersion nor the goodness-of-fit test alone can be used to make inferences about the characteristics of the model. If the null hypothesis is rejected, furthermore, the true type of the distribution will not be discernible. Thus we might turn to a graphical method, perhaps a little less accurate (but certainly well within acceptable limits for most applications), yet considerably simpler and faster. After plotting the empirical sum-percent-curve on Poisson cumulative probability paper (such that the simple Poisson is represented by a perpendicular), in fact, the distribution type may be readily recognized.* Some examples are given later in this article.

2. NOTATION AND SOME PROPERTIES OF DISCRETE DISTRIBUTIONS

The central position of the Poisson distribution can be seen in Table 1. The last line gives a simple indicator for classification as sub-Poisson, Poisson or super-Poisson ** (see also [12] for a discussion of power series). This can also be done by means of Gram-Charlier Type-B series, which is written in terms of the Poisson distribution [14]. The probability that X is equal to c is

$$p[X = c] = \Delta P_c(\lambda) + a_1\Delta^2 P_c(\lambda) + a_2\Delta^3 P_c(\lambda) + \ldots, \qquad (2.1)$$

where a_i are functions of the distribution p(X) which is to be approximated, and

$$\Delta P_c(\lambda) = P_c(\lambda) - P_{c-1}(\lambda) \qquad (2.2)$$

$$\Delta^2 P_c(\lambda) = \Delta P_c(\lambda) - \Delta P_{c-1}(\lambda), \text{ etc.} \qquad (2.3)$$

*The class of compound and generalized Poisson distributions is represented by curves inclined to the right, while the (positive) binomial is versed to the left. Typical curve types and the parameters of these distributions may also be estimated by this method.

**In order to keep notations simple, the binomial, Poisson, and negative binomial distributions have been chosen as representatives of the three classes.

Table 1. Notation and Some Properties of Distributions

	Positive Binomial	Poisson Distribution	Negative Binomial
Frequency Distribution	$b_x(n,p) = \binom{n}{x} p^x q^{n-x}$	$p_x(\lambda) = e^{-\lambda}\left(\frac{\lambda^x}{x!}\right)$	$m_x(k,p) = \binom{k+x-1}{x}\left(\frac{1}{q}\right)^k\left(\frac{p}{q}\right)^x$
Cumulative Distribution	$B_c(n,p) = \sum_{i=0}^{c} b_i(n,p)$	$P_c(\lambda) = \sum_{i=0}^{c} p_i(\lambda)$	$M_c(k,p) = \sum_{i=0}^{c} m_i(k,p)$ (See Note)
Cumulants			
$k_1 = \mu'_1$	np	λ	kp
$k_2 = \mu_2$	np(1-p)	λ	kp(1+p)
k_3	np(1-p)(1-2p)	λ	kp(1+p)(1+2p)
$k = \mu_2/\mu'_1$	$1-p<1$	1	$1+p>1$

NOTE: The letters m and M are in memory of Montmort, who first mentioned this distribution in 1714.

It is well known that

$$\Delta P_c(\lambda) = \frac{\lambda}{c} P_{c-1}(\lambda). \tag{2.4}$$

Thus (2.3) will be written as

$$\Delta^2 P_c(\lambda) = (1 - \frac{c}{\lambda})\Delta P_c(\lambda) \tag{2.5}$$

By comparing moments on both sides of equation (2.1), we obtain a_1. For the first moment we get

$$E(X) = \lambda - a_1$$

Letting $a_1 = 0$ and $E(X) = \lambda$, and equating the second moments, we get

$$E(X^2) = \lambda^2 + \lambda + 2a_2.$$

and thus

$$a_2 = 1/2\left[E(X^2) - E^2(X) - E(X)\right] = 1/2\left[\text{Var}(X) - E(X)\right]. \tag{2.6}$$

For the binomial distribution we have

$$E(X) = np$$

$$\text{Var}(X) = np(1 - p);$$

and thus

$$\lambda = np,$$

$$a_1 = 0,$$

$$a_2 = -1/2np^2 \quad \text{and}$$

$$b_c(n,p) = \Delta P_c(np) - 1/2np^2 \cdot \Delta^3 P_c(np) + R \tag{2.7}$$

with

$$R = \frac{e^{2\lambda^2/n}}{n^2\pi}\left[1/8I_4(\lambda) + 1/3I_3(\lambda)(1 - \frac{2\lambda}{n})^{-1} \cdot \exp\frac{8\lambda^3}{3\lambda^2(1-2\lambda/n)}\right]$$

and

$$I_j(\lambda) = \lambda^j \int_0^\pi t^j \exp\left[-2\lambda \sin^2 1/2t\right]dt.$$

Further we have

$$B_c(n,p) = P_c(np) - 1/2np^2\Delta^2 P_c(np) + R'$$

and using (2.5):

$$B_c(n,p) = P_c(np) - 1/2(np - c)\Delta P_c(np) + R', \qquad (2.8)$$

where $np = \lambda$.
In order to get the corresponding formulae for the negative binomial distribution p must be negative and we put -k instead of n, so the sign of the second term in (2.8) is reversed and we get

$$M_c(k,p) = P_c(kp) + 1/2\ p(kp - c)\ \Delta P_c(kp) + R''\ . \qquad (2.9)$$

Tables for maximal deviation in (2.8) and (2.9) for various p and n, p and k are given elsewhere ([7]). There we see that this approximation is good for small p's. In the cases for $c = np$ and $c = kp$, respectively, we have very good approximations to the related Poisson term.

3. CONSTRUCTION OF THE POISSON PROBABILITY PAPER

Figure 1 is a modification of one of Campbell's diagrams (see [2]), in which the ordinate has the percent-scores of normal distribution. The approximation that follows was used for the calculations:

$$P_c(\lambda) = (2\pi)^{-1/2} \int_{-\infty}^{y} \exp(-1/2t^2)dt + 1/6(2\pi\lambda)^{-1/2}(1-y^2)\ \cdot$$

$$\exp(-1/2y^2) + \delta,$$

where

$$y = \lambda^{-1/2}(c - \lambda + 1/2)$$

and $|\delta| < 0.076\lambda^{-1} + 0.043\lambda^{-3/2} + 0.13\lambda^{-2}$ (from [3]).

Because the transformed Poisson random variables have approximately normal distribution with mean $\sqrt{\lambda}$ and stable variance 1/4 (see [5]), the abscissa has a square root transformation.
In Figure 1 the curves for

$$\text{Prob}\quad X \le c\ = P_c(\lambda) \qquad c = 0,1,2,\ldots.$$

are plotted so that the sum-percent curve of the Poisson lies on a perpendicular line. On the abscissa we find an estimate of λ. Formulas (2.8) and (2.9) show that the differences between

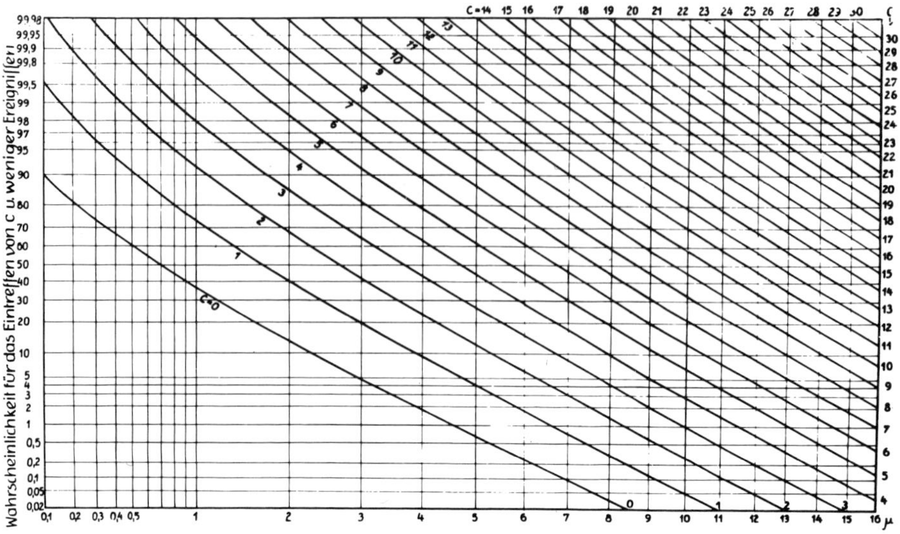

Figure 1. Poisson Cumulative Probability Paper

the binomial (or negative binomial) and the corresponding Poisson terms mainly depend upon c, when p, n (and k) are given. So we have

$$B_c(n,p) \begin{cases} <P_c(n,p) & \text{for } c < np \\ = P_c(n,p) & \text{for } c = np \\ >P_c(n,p) & \text{for } c > np \end{cases} \qquad (3.1)$$

Joining the points of the sum-percents of the binomial distribution, we get the curves of Figure 2. Their slope increases with increasing $q = 1 - p$.

A transparency of Fig. 2 enables us to compare the empirical curve with the theoretical ones and estimate np, as well as q. Using tables in [8] we derive Figure 3 with curves of the negative binomial, which have slopes in the opposite direction.

For negative binomial distribution the inequalities corresponding

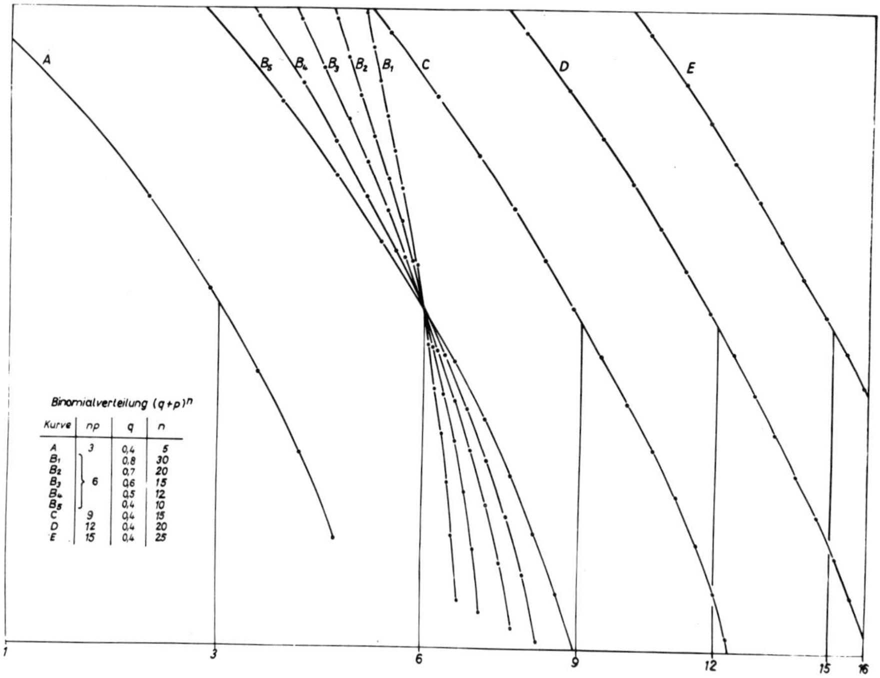

Figure 2. Binomial Distribution with np = 3, 6, 9, 12, 15 and n = 5, 10, 12, 15, 20, 25, 30.

to (3.1) are

$$M_c(k,p) \begin{cases} > P_c(kp) & \text{for } c < kp \\ = P_c(kp) & \text{for } c = kp \\ < P_c(kp) & \text{for } c > kp \end{cases} \tag{3.2}$$

Similar transparencies have been drawn using tables in [7] and [9], for the Neyman distributions of Type A and Type $n \to \infty$ (Figures 4 and 5).

The values of the zero-classes of the various distributions have the following order:
P_0(Neyman A) > P_0(Neyman B) > ... > P_0(Neyman $n \to \infty$) > M_0 > P_0(Poisson) > B_0. The tables in [15] provide data for curves of the logarithmic distribution. Keeping in mind these typical

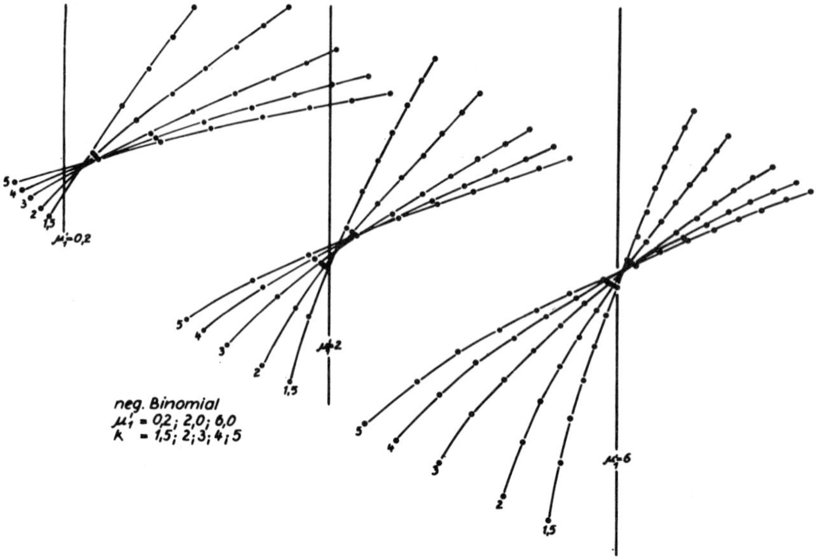

Figure 3. Negative Binomial Distribution with $\mu_1' = 0.2, 2.0, 6.0$, and $k = \mu_2/\mu_1' = 1.5, 2, 3, 4, 5$

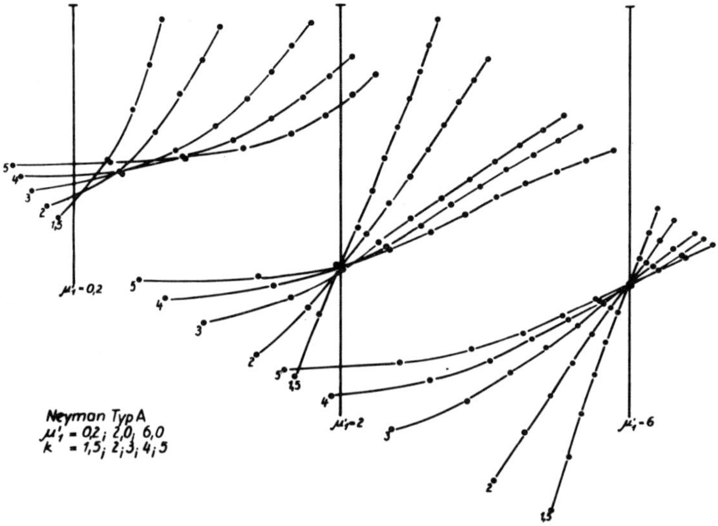

Figure 4. Neyman Distribution Type A with the Same Parameters as in Fig. 3

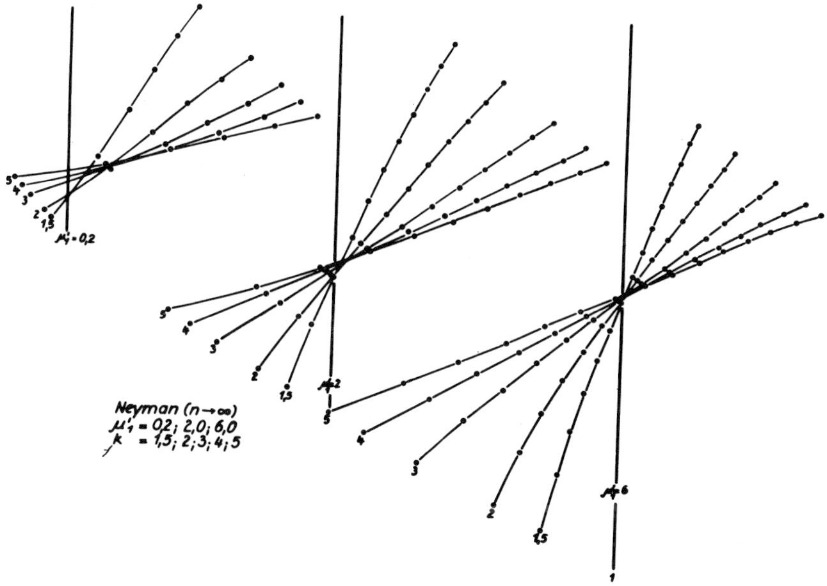

Figure 5. Neyman Distribution Type n→∞ with the Same Parameters as in Fig. 3

curves, we can decide which transparency is to be used. On the transparencies for each of these sets of curves the corresponding perpendicular Poisson curve with the same mean is given. This fact and two-fold interpolation between λ-values of the abscissa and the corresponding c-curves may be used in finding the mean of the distribution. From the slope k, an estimate of μ_2 is derived using the relation $\mu_2 = \mu_1' k$.

4. AN EXAMPLE

In Table 2 we have data on the distribution of Bacillus Mycoides and the sum-percent curves are given in Fig. 6. The distribution of the number of colonies is Poisson and the distribution of the number of germs in a colony has a logarithmic distribution so the distribution of the total number of germs on a plate is negative binomial. The estimates of the parameters are mean = 0.217, q = 2.089, k = 0.196, and variance = 0.908. Further examples including one with false enumeration in the zero class are given in [7].

Table 2. Germ Distribution of Bacillus Mycoides

no. of germs in unit squares	frequency	sum-percents
x	f	S%
0	1419	86.5
1	149	95.6
2	42	98.2
3	16	99.1
4	6	99.5
5	5	99.8
6	1	99.9
7	2	100.0
Sum	1640	

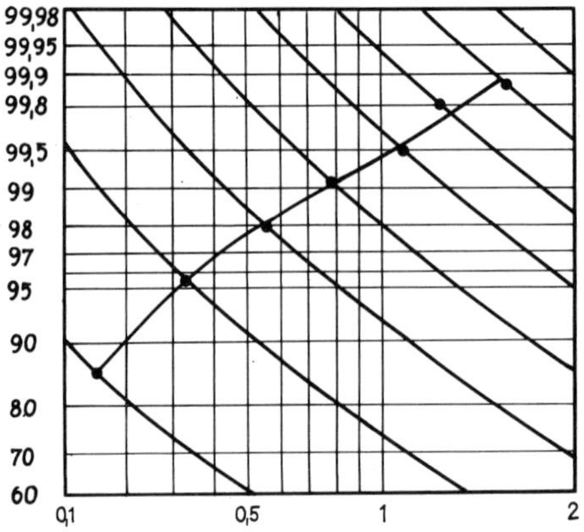

Figure 6. Sum-Percent Curves of Bacillus Mycoides in Cumulative Poisson Probability Paper

5. A Sampling Study and Comparison with Other Methods

240 Random samples of size $N = 10, 50, 100$ and 144 have been drawn from simple Poisson distribution with $\lambda = 0.2, 2.0$ and 6.0; and another 240 from the negative binomial with $kp = 0.2, 2.0, 6.0$, and $k = 1.5$. When plotting this information, we find that the critical slope k_α at the level α can be calculated from $k_\alpha = \chi^2_{\alpha;N-1}/N-1$. Some values for $\alpha = 0.05$ are given in Table 3:

Table 3:
Critical Slope for $\alpha = 0.05$, and $N = 10, 50, 100, 144$ and 1000

N	$k_{0.05}$
10	1.83
50	1.35
100	1.24
144	1.20
1000	1.07

The null hypothesis H_0: simple Poisson with $\hat{\lambda} = 0.2, 2.0, 6.0$, respectively has been tested at the 5 percent level with the variance test (index of dispersion) ([4]), with the likelihood ratio of Rao and Chakravarti [13], and using the graphical method. Table 4 gives the number of cases in which H_0 was rejected. In the first column (P) are samples from simple Poisson; in the second one (M), those from negative binomial. We remark that for $\mu_1' = 0.2$, the likelihood ratio test fails to detect negative binomial. Thus the quick graphical method is even better for such cases.

6. GRAPHICAL DETERMINATION OF 100(1 - α)% CONFIDENCE INTERVALS

First, the lower Poisson confidence limit λ_ℓ at the level α is determined by

$$\sum_{i=0}^{c-1} e^{-\lambda_\ell} \frac{\lambda_\ell^i}{i!} = 1 - \frac{\alpha}{2},$$

and the upper limit by

Table 4. Number of Cases in which H_0 (Simple Poisson) was Rejected at the Level $\alpha = 0.05$

μ'_1	N	Index of Dispersion		Likelihood Test		Graphical Method	
		P	M	P	M	P	M
0.2	10	0	3	0	0	0	0
	50	1	10	0	0	1	11
	100	0	13	0	0	0	14
	144	1	14	0	0	1	15
2.0	10	1	5	2	7	2	7
	50	1	12	3	13	1	14
	100	0	18	8	19	0	19
	144	0	19	6	20	0	20
6.0	10	1	4	3	5	3	8
	50	0	10	0	9	0	11
	100	2	19	2	19	2	19
	144	1	19	1	19	1	19

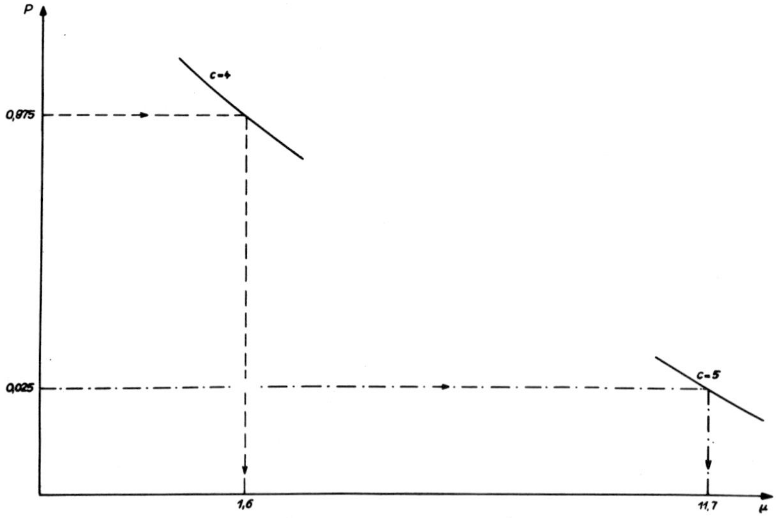

Figure 7. Determination of the 95% Confidence Interval for Poisson Variable c = 5

$$\sum_{i=0}^{c} e^{-\lambda_u} \frac{\lambda_\ell^i}{i!} = \frac{\alpha}{2} .$$

For c = 5 and α = 0.05 we go from the ordinate at 97.5% to the intercept with c = 4 and find on the abscissa λ_1 = 1.6. From the ordinate 2.5%, we go to c = 5, finding λ_u = 11.7. See Fig. 7. Next, the confidence limit for the negative binomial distribution with known k (e.g., k = 1.5) and c (say, 5) is found by moving the transparency and using the line for k = 1.5. Thus we find the limits 1.3 and 12.8; see Fig. 8. In many cases Figure 1 and the transparencies may be used when tables of the distributions are not available.

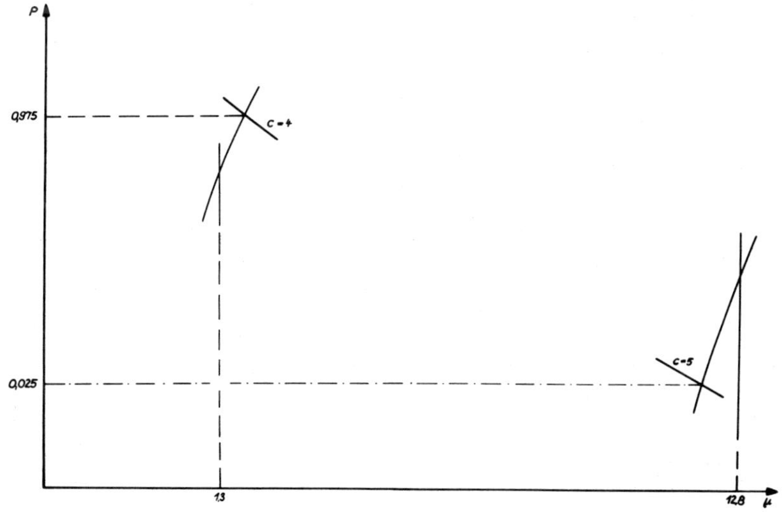

Figure 8. Determination of the 95% Confidence Interval for Negative Binomial Distribution, k = 1.5 and c = 5.

REFERENCES

[1] Bliss, C. J. 1953. Fitting the negative binomial distribution to biological data. Biometrics 9:176-96.

[2] Campbell, G. A. 1923. Probability curves showing Poisson's exponential summation. Bell System Technical Journal 2:95-113.

[3] Cheng, T. T. 1949. The normal approximation to the Poisson distribution and a proof of a conjecture of Ramanujan. Bull. Amer. Math. Soc. 55:396-401.

[4] Cochran, W. G. 1954. Some methods for strengthening the common χ^2 tests. Biometrics 10:417-51.

[5] Grimm, H. 1960. Transformation von Zufallsvariablen. Biometrische Zeitschrift 2:164-82.

[6] ————. 1961. Quelques problèmes de numérations bactériennes. Biometrie-Praximétrie (Brüssel) 2:1-18.

[7] ————. 1962a. Prüf und Schatzemethoden bei einfachen und zusammengesetzten Poisson-Verteilungen. Dissertation Universität Leipzig (unpublished).

[8] ————. 1962b. Tafeln der negativen Binomialverteilung. Biometrische Zeitschrift 4:239-62.

[9] ————. 1964. Tafeln der Neyman-Verteilung Typ A. Biometrische Zeitschrift 6:10-23.

[10] ————. 1964. Über einige stochastische Prozesse, die zum Auftellen mathematischer Modelle geeignet sind. Abh. D.A.W. Klasse Math. Phys. Techn. No. 4, 65-8.

[11] Haight, F. 1967. Handbook of the Poisson Distribution. John Wiley & Sons, New York.

[12] Patil, G. P. 1962. Certain properties of generalized power series distribution. Ann. Inst. Statist. Math. 14:179-82.

[13] Rao, C. R., and Chakravarti, I. M. 1956. Some small sample tests of significance for a Poisson distribution. Biometrics 12:264-82.

[14] Raff, M. S. 1956. On approximating the point binomial. Journ. Amer. Statist. Assoc. 51:293-303.

[15] Williamson, E., and Bretherton, M. H. 1964. Tables of the logarithmic series distribution. Ann. Math. Statist. 35:284-94.

SOME EXACT METHODS OF INFERENCE APPLIED TO DISCRETE DISTRIBUTIONS

D. A. SPROTT*
University of Waterloo

SUMMARY

Applications of some exact methods of inference to discrete probability distributions are described, and the use of likelihoods and likelihood contours is illustrated and investigated for problems of one and two parameters. Maximum relative likelihoods and conditional likelihoods are defined and illustrated as methods of eliminating unwanted parameters. Exact goodness-of-fit tests by enumeration, discussed in this article, can be performed on discrete distributions when the parameters can be eliminated by sufficient statistics. When this is not the case, a test of significance (such as chi-square) can be simulated on a computer to give an estimated significance level, along with the upper and lower bounds. The above methods are all exact in the sense that, unlike asymptotic methods, they do not depend on large sample sizes for their validity. Although the above methods entail much more tedious computations, access to high-speed computers now makes them feasible.

* With a grant under the auspices of the National Research Council.

1. INTRODUCTION

This paper exemplifies some exact methods of inference applied to discrete distributions. An exact (as opposed to an asymptotic) method of inference is a method which does not use mathematical approximation (depending on the sample size) for its validity. Examples of asymptotic procedures are the theory of maximum likelihood estimation and the Chi-square goodness-of-fit test; exact alternatives for these methods are considered herein.

In review, the theory of maximum likelihood states that in large samples the maximum likelihood estimate $\hat{\theta}$ is normally distributed about θ with a standard error usually estimated as $V(\hat{\theta})$. This allows approximate probability intervals (fiducial or Bayes) for θ, as well as confidence intervals, to be constructed. The Chi-square goodness-of-fit test is based on the usual χ^2 departure criterion having the χ^2 distribution in large samples.

Sections 2, 3, and 4 exemplify the use of the likelihood function as a method of summarizing the data and of setting up plausibility intervals for the parameters (i.e., interval estimation). Since the likelihood function is defined irrespective of the sample size, no asymptotic assumptions are necessary. However, it is necessary to measure the plausibility in terms of likelihood rather than probability or confidence levels. One has then to decide whether exact statements of likelihood are preferable to approximate statements of probability, keeping in mind that the degree of approximation is not usually known.

Section 2 deals with problems involving single parameters and compares the exact likelihood approach with the maximum likelihood approach. Use of maximum likelihood essentially assumes the sample is large enough to yield the normal likelihood function as a good approximation. It will be seen that this is often not the case, but that a transformation of the parameter sometimes improves the situation.

Sections 3 and 4 deal with multiparameter likelihoods. The difficulty, then, is the elimination of unwanted parameters in order to make a statement about a single parameter of interest. Two methods for doing this, maximum relative likelihoods and conditional likelihoods, are described and compared. Similar ideas are discussed in [5], [1], [16], and [17].

Sections 5 and 6 deal with goodness-of-fit tests. Specifically, 5 illustrates exact tests by enumeration when unknown parameters can be eliminated by sufficient statistics; Section 6 exemplifies the simulation of tests so that an estimate of the significance level can be obtained.

Finally the main point to be illustrated in this paper is that, whereas asymptotic methods of inference previously had to be employed, the accessibility of high speed computers now makes the above exact methods feasible.

2. SINGLE PARAMETER LIKELIHOODS

2.1. Likelihood and Relative Likelihood.

Suppose the observations $x_1, x_2, \ldots, x_n$ are a random sample from the distribution $p(x;\theta)$. The distribution of the observations is then $p(x_1, x_2, \ldots, x_n;\theta) = \pi p(x_i;\theta)$. The likelihood function of θ is proportional to

$$L(\theta) = \prod_{i=1}^{n} p(x_i;\theta).$$

The ratio $L(\theta_1)/L(\theta_2)$ can be taken as a measure of the plausibility of θ_1 vs θ_2 in the light of the observations $x_1, \ldots, x_n$. For instance if $L(\theta_1)/L(\theta_2) = 2$, then θ_1 is twice as plausible as θ_2 in the sense that the actual observations are twice as probable if θ_1 is true than if θ_2 is true.

It is convenient to standardize the likelihood function with respect to its maximum $\sup L(\theta) = L(\hat{\theta})$, where $\hat{\theta}$ is the maximum likelihood estimate of θ. This in effect ranks the plausibilities of values of θ with respect to the most plausible value $\hat{\theta}$. The resulting function, $R(\theta) = L(\theta)/L(\hat{\theta})$, is the relative likelihood function of θ. Like a probability, $R(\theta)$ varies between 0 and 1. Unlike probabilities, however, they cannot be combined to form likelihoods of compound events. Thus unwanted variables cannot be eliminated by integration.

It should be noted that statements of likelihood have a simple frequency interpretation. $R(\theta_0) = 0.1$ means that samples equivalent to the observed one are 10 times as probable if $\theta = \hat{\theta}$ than if $\theta = \theta_0$. Such a statement could be verified empirically on a computer to any degree of accuracy by actually sampling the respective populations.

Of course it is possible to standardize the likelihood in other ways, such as with respect to its average. This would give a function

$$L(\theta)\Big/\int_{-\infty}^{\infty} L(\theta)d\theta,$$

which integrates to unity like a probability distribution.

2.2 Relation to Large Sample Theory.

The general theory of maximum likelihood states that asymptotically $\hat{\theta}$ has the normal distribution $N\ \theta,\sigma^2(\theta)$, where the asymptotic variance $\sigma^2(\theta)$ is

$$\sigma^2(\theta) = -1\Big/E\left[\frac{\partial^2 \log L(\theta)}{\partial\theta^2}\right]. \tag{2.1}$$

This variance is usually estimated by $\sigma^2(\hat{\theta})$, or by

$$-1\Big/\frac{\partial^2 \log L(\theta)}{\partial\hat{\theta}^2}\ . \tag{2.2}$$

This would imply that in large samples the observed likelihood can usually be approximated by a normal likelihood

$$R_N(\theta) = \exp\left[-\frac{1}{2\sigma^2}(\theta-\hat{\theta})^2\right], \tag{2.3}$$

where σ^2 is the estimated variance.

This is a simplification since the information in the sample can be approximated by a point estimate and a variance. Approximate statements of probability (fiducial or Bayesian) or confidence intervals can be based on this in the usual way.

Example: Dilution Series. This example has been discussed in some detail in [6] and [3]. Suppose the average density of organisms in solution is n organisms per unit volume. Dilute a unit volume by factors $1,a,a^2,\ldots,a^k$ to give a series of solutions of average densities $n,n/a,n/a^2,\ldots,n/a^k$. Innoculate s plates with a solution of density n/a^i for each $i = 0,1,\ldots,k$, and let y_i be the number of sterile plates observed out of the s at level i. Assuming a Poisson distribution of organisms, then, the probability of the observations $y_0,y_1,\ldots,y_k$ is

$$\prod_{i=0}^{k}\binom{s}{y_i}p_i^{\,y_i}(1-p_i)^{s-y_i},$$

where $p_i = \exp(-\lambda_i)$, $\lambda_i = n/a^i$; the likelihood of the unknown parameter n is thus proportional to

$$L = \prod_{i=0}^{k} p_i^{y_i}(1-p_i)^{s-y_i}.$$

[6] Features tables from which a slightly inefficient estimate of n can be obtained with computational ease. However since high-speed computers are now fairly accessible, such approximate methods are not so important as the maximum likelihood estimate can be obtained and the whole likelihood of n tabulated in a few seconds.

The maximum likelihood estimate $\hat{n}$ satisfies the equation

$$g = \frac{\partial \log L}{\partial n} = \sum \frac{sp_i - y_i}{(1-p_i)a^i} = 0. \qquad (2.4)$$

Also

$$g' = \frac{\partial^2 \log L}{\partial n^2} = -\sum \frac{(s-y_i)p_i}{(1-p_i)^2 a^{2i}} .$$

Equation (2.4) can be solved iteratively by Newton's method: n_α converges to $\hat{n}$ where $n_{\alpha+1} = n_\alpha - g/g'$. The relative likelihood

$$R(n) = \pi p_i^{y_i}(1-p_i)^{s-y_i} \Big/ \pi \hat{p}_i^{y_i}\left(1-\hat{p}_i\right)^{s-y_i} , \qquad (2.5)$$

$$\hat{p}_i = \exp(-\hat{n}/a^i),$$

can be then tabulated. Computer programs can quite simply be written to do this. In the example of [6], in particular, $a = 2$, $s = 5$, $k = 9$ and $y_i(i = 0,9)$ are 0,0,0,0,1,2,3,3,5,5. Starting with an initial value $n = 30.4$, the solution to (4) was found in 4 seconds on an IBM 7090 computer: $\hat{n} = 30.649790$.

The relative likelihood (2.5) can then be calculated for values of 25n between 250 and 1600 in steps of 25 (the factor 25 for converting the result into units of organisms/gram), as shown in Figure 1. Execution time on the IBM 7040 is about 10 seconds for this computation.

The estimated asymptotic variance of $25\hat{n}$ calculated from (2.2) is

$$25^2/(0.01225) = 51020.4081,$$

from which the approximating normal likelihood R_n (25n) given by (2.3) can be calculated. A comparison with R(25n) in Figure 1 reveals that the agreement is not very close. Therefore the above use of large sample approximations to set up intervals of uncertainty about n would be misleading. For instance, values of 25n in the range (400, 1400) have relative likelihood of 10% or more and so are fairly plausible. Using the normal approximation, the 10% range is (180, 1250).

2.3 Functional Invariance and Transformations.

An important property of inferences based on the likelihood function is functional invariance. That is, any statement of plausibility about a parameter θ can be converted into a corresponding statement about any single-valued functions of θ merely by substitution. This allows the effects of transformations of the parameter to be investigated easily and quickly using a high-speed computer. (Other uses of functional invariance are illustrated in [17].)

Frequently a transformation of the parameter improves the normal approximation. In Example 1, above, the effect of using log 25n can be ascertained, in fact, merely by inserting a single instruction in the program for calculating the relative likelihood. Then $\mu = \log 25n$ can be tabulated along with R(25n), so that R(25n) can be plotted on the logarithmic scale. The result is shown in Figure 2, in which $R(\mu)$ is plotted. The maximum likelihood estimate of log 25n is $\hat{\mu} = \log 25\hat{n}$, and the asymptotic variance of $\log 25\hat{n}$ can be estimated as

$$\mathrm{var}(\hat{\mu}) = \mathrm{var}(25\hat{n})\left(\frac{d\log 25n}{d\hat{n}}\right)^2 = 0.086592 .$$

From this the large sample normal approximation $R_N(\mu)$ can be calculated. (See Figure 2 for a comparison with $R(\mu)$. It can be seen that the agreement is much closer than in Figure 1.) Using the normal approximation, values of log 25n in the range (6.02, 7.28) have at least 10% relative likelihood. This yields the range (411, 1450) for 25n which is in closer agreement with the range (400, 1400) obtained from the exact likelihood in Section 2.2.

[6] Actually used the parameter μ in setting up intervals of uncertainty for 25n. They obtain 5% fiducial limits for 25n of (407, 1440), which compares favorably with the likelihood results. Applications of maximum likelihood theory directly to n itself would yield different and somewhat misleading results, owing to the non-normal likelihood of n as shown in Section 2.2.

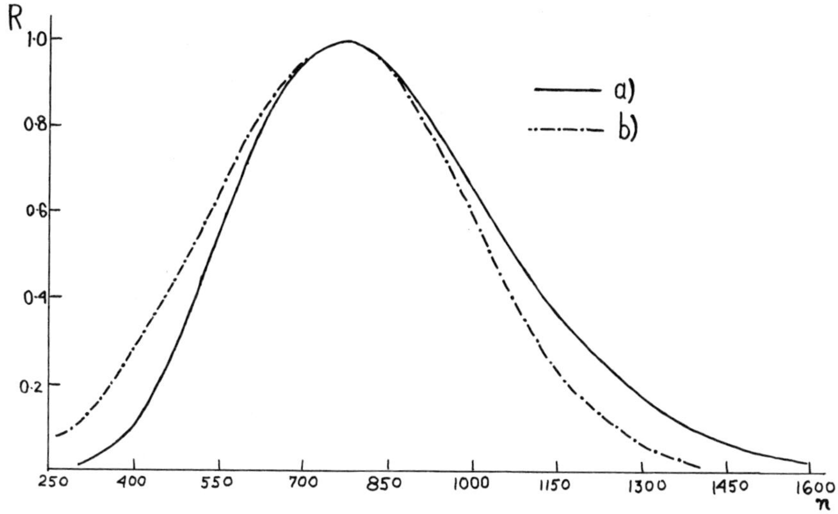

Figure 1. Relative likelihood $R(\nu)$ of ν and normal approximation $R_N(\nu)$ for the dilution series data of Example 1, $\nu = 25n$.

(a) $R(\nu)$, (b) $R_n(\nu)$.

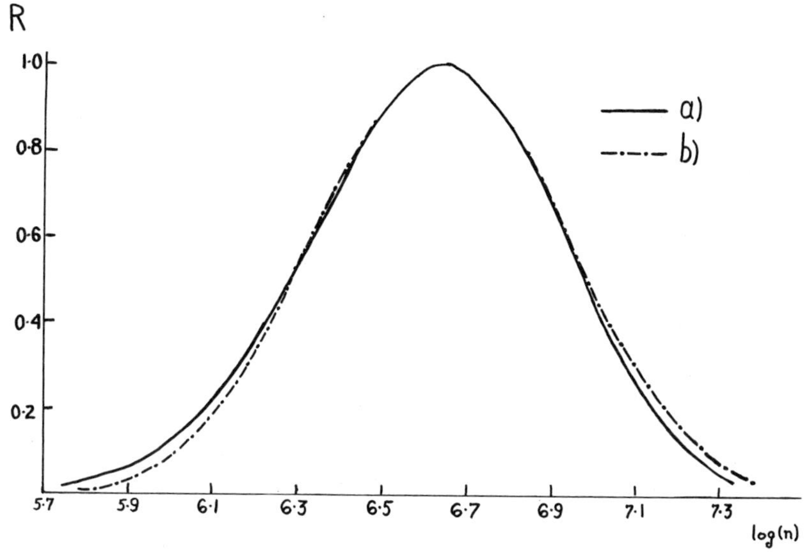

Figure 2: Relative likelihood $R(\mu)$ of $\mu = \log \nu$ and normal approximation $R_N(\mu)$ for the dilution series data, $\nu = 25n$.

(a) $R(\mu)$, (b) $R_N(\mu)$.

3. MULTIPARAMETER LIKELIHOODS

3.1. Likelihood Contours.

Likelihoods containing more than one parameter present certain difficulties. If there are many parameters, it becomes necessary to eliminate the unwanted ones so that inferences can be made about the parameter of interest in the absence of knowledge of the unwanted parameters. If there are only two parameters, it is possible to plot a graph of likelihood contours, as illustrated in Example 2.

Example 2: Sprott [15] fitted a logarithmic binomial distribution with generating function

$$G(\theta_1,\theta_2,s) = \frac{\log\left[1-\theta_1(1-\theta_2+\theta_2 s)^2\right]}{\log(1-\theta_1)},\quad 0 \le \theta \text{ and } \theta_2 \le 1,$$

to some data found in [12]. The maximum likelihood estimates were $\hat{\theta}_1 = 0.9816$, $\hat{\theta}_2 = 0.0717$, very close to the extreme ends of the range, so that it might be expected that the large sample normal approximation will not be very accurate. From the likelihood contours shown in Figure 3, it can be seen that they depart substantially from the elliptical form implied by the normal distribution. Values of θ_2 in the range (0.025, 0.240) have relative likelihood greater than 1%; values of θ_1 in the range (0.896, 0.995) have relative likelihood greater than 1%. Note also that θ_1 is being estimated more precisely than θ_2 and that, from the contours, knowledge of one of the parameters affects the inference concerning the other.

Barnard (personal communication) suggested a transformation of the form

$$\alpha_1 = \log(a\theta_1 - b\theta_2)$$

$$\alpha_2 = c\theta_1 + d\theta_2,$$

which might be used, as in section 2.3, to improve the normal approximation in Example 2. (As yet this possibility has not been checked.)

3.2. Maximum Relative Likelihood.

As indicated previously, the elimination of unwanted parameters from the likelihood presents some difficulties. This is the advantage in having a probability distribution rather than a

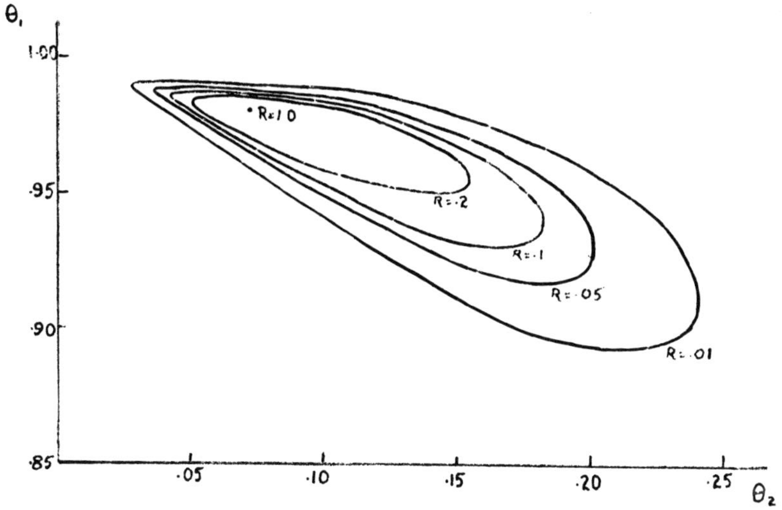

Figure 3. Relative likelihood contours of θ_1, θ_2, for the data of Example 2.

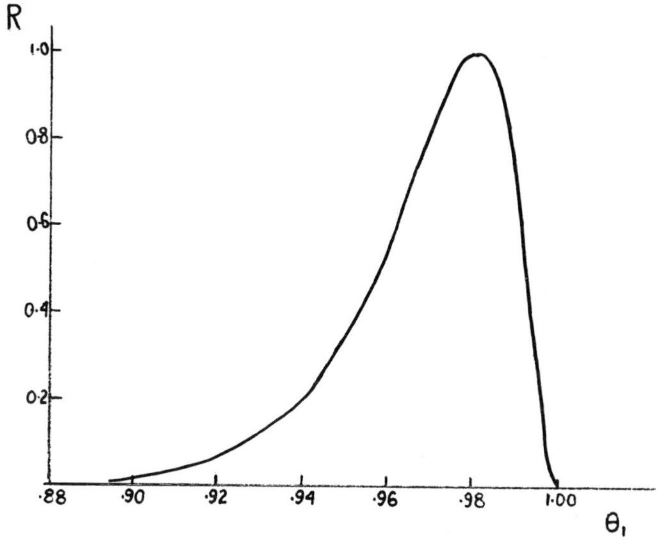

Figure 4. Maximum relative likelihood of θ_1 for the data of Example 2.

likelihood, since unwanted parameters can be removed from a probability distribution by integration.

One possible way to remove θ_2 from the relative likelihood $R(\theta_1, \theta_2)$ is to replace θ_2 by its maximum likelihood estimate θ_2' (for the specified θ_1). It is noteworthy that, for a fixed θ_1, the maximum likelihood estimate of θ_2 will probably be a function of θ_1, $\theta_2' = \theta_2'(\theta_1)$. This gives the maximum relative likelihood of θ_1

$$\sup_{\theta_2} R(\theta_1, \theta_2) = R\left[\theta_1, \theta_2'(\theta_1)\right] . \tag{3.1}$$

That is, whatever be the value of θ_2, the relative likelihood of θ_1 cannot exceed (3.1).

The maximum relative likelihood (3.1) of θ_1 in Example 2 is plotted in Figure 4. Its non-normal shape is apparent, and again the effect of various transformations can readily be ascertained. In Figure 5 the maximum relative likelihood of

$$\phi_2 = 1/2\log\left(\frac{\theta_1}{1-\ _1}\right)$$

and the likelihood arising from its normal approximation are plotted. Although the agreement is still not very good, the maximum relative likelihood of ϕ_2 is much closer to normality.

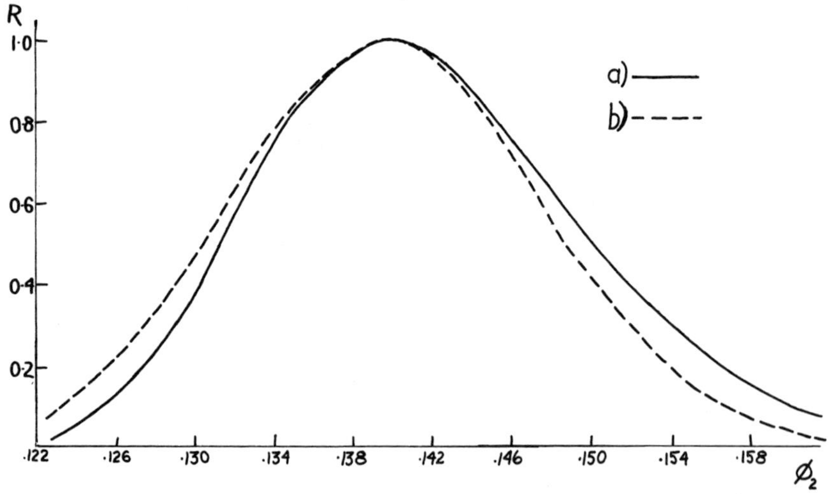

Figure 5. (a) Maximum relative likelihood of $\phi_2 = \frac{1}{2}\log\frac{\theta_1}{1-\theta_1}$ and (b) the normal approximation for the data of Example 2.

Consequently, it would seem that the only accurate statements of uncertainty that can be made about θ_1 or θ_2 in Example 2 must be given in terms of likelihood. (For other examples of the uses of likelihood contours and maximum relative likelihoods, see [17].)

3.3. Conditional Likelihood.

A defect of maximum relative likelihood is that the nuisance parameters are assumed to be equal to their maximum likelihood estimates. This, in effect, assumes they are known exactly; thus the loss of precision due to uncertainty concerning these nuisance parameters is not reflected in the maximum of the relative likelihood. If many parameters are removed in this way, the loss of precision could be serious; and the maximum relative likelihood, consequently, would be misleading. An obvious example of this dilemma is found in the estimation of the variance σ^2 in the normal distribution when k regression parameters have been estimated. The maximum likelihood estimate of σ^2 and corresponding maximum relative likelihood of σ^2 do not take into account the k degrees of freedom lost in estimating the regression parameters. If k is close to n, σ^2 should be estimated with very low precision even if n is large; and the maximum relative likelihood is therefore misleading. This effect is usually not very serious unless the number of parameters being estimated and eliminated is large compared to the total number of observations.

Conditional likelihood overcomes this defect by eliminating the unwanted parameters by the use of sufficient statistics. That is, if a statistic T is sufficient for θ_2 when θ_1 is known, the conditional distribution of the sample, given T, must be independent of θ_2. In this way the unwanted parameter θ_2 has been eliminated by conditioning on a statistic sufficient for it. The conditional distribution

$$p(x_1,\ldots,x_n;\ \theta_1|T) = p(x_1,\ldots,x_n;\ \theta_1,\theta_2)/p(T;\ \theta_1,\theta_2),$$

being a function of θ_1 only, is available for inferences concerning θ_1 in the absence of knowledge of θ_2.

Examples of a similar procedure using ancillary statistics (marginal likelihood) were first given in [7] and [8]. Both the use of ancillary statistics and sufficient statistics for the elimination of unwanted parameters from a likelihood function are discussed in some detail in [8] and [11].

Conditional likelihoods thus arise from factoring the distribution function of the observations

$$p(x_1,\ldots,x_n;\ \theta_1,\theta_2) = p(T;\ \theta_1,\theta_2)\ p(x_1,\ldots,x_n;\ \theta_1|T).$$

Here T is sufficient for θ_2 if θ_1 is known. The second factor is available for inferences about θ_1 in the absence of knowledge of θ_2 and can be called the conditional likelihood of θ_1. An example of this is the estimation of k in the negative binomial distribution (consider r distributions) with a common k:

$$p(i) = \binom{k+i-1}{k-1} p_j^{\,i} q_j^{\,-k-i},\quad j = 1,2,\ldots,r,$$

where $q_j = 1+p_j$. Further, suppose i is observed with frequency a_{ij} in the j-th distribution. Then the probability of the observations a_{ij} is

$$\prod_{i,j} \binom{k+i-1}{k-1}^{a_{ij}} (p_j)^{ia_{ij}} (q_j)^{(-k-i)a_{ij}}$$

$$= \prod_i \binom{k+i-1}{k-1}^{a_{i*}} \left(\prod_j p_j\right)^{T_j} (q_j)^{-ka_{*j}-T_j}, \tag{3.2}$$

where

$$a_{i*} = \sum_j a_{ij},\ \text{and}\ a_{*j} = \sum_i a_{ij}$$

and

$$T_j = \sum_i i a_{ij}.$$

Since T_j has the negative binomial distribution with parameters ka_{*j} and p_j, and the T_j are mutually independent, the joint distribution of the T_j is

$$p(T_1,\ldots,T_r) = \prod_j \binom{ka_{*j} + T_j-1}{ka_{*j}-1} (p_j)^{T_j} (q_j)^{-ka_{*j}-T_j}.$$

The conditional distribution of the observations, given $T_1,T_2,\ldots,T_r$, is

$$p(a_{ij}|T_i) = \prod_i \binom{k+i-1}{k-1}^{a_{i*}} \Big/ \prod_j \binom{ka_{*j}+T_j-1}{ka_{*j}-1}, \tag{3.3}$$

these values $\hat{k}_{ij}$. The advantage of this procedure is that it is unnecessary to recalculate new estimates of k_{ij} for each set of values of the remaining parameters.

The observations can be represented by $a_{ij\alpha}$, the frequency with which α occurs in the population (i,j). The distribution of the observations is then

$$p(a_{ij\alpha}) = \prod_{ij\alpha} \left[\binom{k_{ij}+\alpha-1}{k_{ij}-1} p_{ij}^{\alpha}(1+p_{ij})^{-k_{ij}-\alpha} \right]^{a_{ij\alpha}}$$

The sample mean

$$\bar{a}_{ij} = \frac{1}{n_{ij}} \sum_{\alpha} \alpha_{ij\alpha}$$

is the maximum likelihood estimate of $\eta_{ij} = k_{ij}p_{ij}$, where η_{ij} is the total number of observations from the population (i,j),

$$n_{ij} = \sum_{\alpha} a_{ij\alpha} \text{ .}$$

The logarithm of the likelihood is

$$L = \sum_{ij\alpha} a_{ij\alpha} \log \binom{k_{ij}+ \alpha-1}{k_{ij}-1} + \sum_{ij} \left[T_{ij} \log \frac{\eta_{ij}}{k_{ij}} - (n_{ij}k_{ij}+T_{ij}) \log \left(1+\frac{\eta_{ij}}{k_{ij}}\right) \right] , \tag{4.2}$$

where

$$T_{ij} = \sum_{\alpha} \alpha a_{ij\alpha} = n_{ij}\bar{a}_{ij} ,$$

and

$$\eta_{ij} = k_{ij}p_{ij} \text{ .}$$

Thus the rc maximum likelihood estimates $\hat{\eta}_{ij}$, $\hat{\rho}_i$, $\hat{\beta}_k$, $\hat{\gamma}_{ij}$, $\hat{\mu}$ satisfy the equations

$$\hat{\eta}_{ij} = \bar{a}_{ij} = \hat{\mu} + \hat{\rho}_i + \hat{\beta}_j + \hat{\gamma}_{ij},$$

$$\sum_i \hat{\rho}_i = \sum_j \hat{\beta}_i = \sum_i \hat{\gamma}_{ij} = \sum_j \hat{\gamma}_{ij} = 0.$$

The solutions are the usual

so that $T_1, \ldots, T_r$ are sufficient for the p's. Thus the conditional likelihood of k is proportional to (3.3). This can be standardized with respect to its maximum to obtain the relative conditional likelihood and this can be compared with maximum relative likelihood obtained from (3.2) by replacing all p_j by their maximum likelihood estimate T_j/ka_{*j}. The result is proportional to

$$\prod_{i,j} \binom{k+i-1}{k-1}^{a_{i*}} \left(T_j\right)^{T_j} \left(ka_{*j}\right)^{ka_{*j}} \left(ka_{*j}+T_j\right)^{ka_{*j}+T_j} \quad . \tag{3.4}$$

Functions (3.3) and (3.4) usually produce very similar results; they will differ most when the number of r of p_j''s being estimated is a large fraction of the total number of observations.

<u>Example 3</u>: Ten negative binomial distributions with $k = 2$ were sampled on a computer. The sample sizes a_{*j}, $j = 1,2,\ldots,10$, were deliberately made small; they were 3,3,4,5,6,3,4,5,4,6, giving the total number of observations as 43, compared to ten parameters p_j to be estimated. The conditional relative likelihood arising from (3.3) is compared with the maximum relative likelihood arising from (3.4) in Figure 6. Through a search technique or an iterative procedure, the maximum of (3.3) and of (3.4) can be calculated fairly quickly. It can be seen that there is an appreciable difference and that the maximum relative likelihood in this case is somewhat misleading. Two other examples of this kind were generated on a computer with very similar results, and in each case the conditional likelihood was more closely centred on the true value $k = 2$.

4. ANALYSIS OF UNTRANSFORMED DATA FROM CONTAGIOUS DISTRIBUTIONS

Hinz and Gurland [10] have considered discrete distributions in a two-way classification where the means η_{ij} can be written

$$\eta_{ij} = \mu+\rho_i+\beta_j+\gamma_{ij}, \tag{4.1}$$

$$\sum_i \rho_i = \sum_j \beta_j = \sum_i \gamma_{ij} = \sum_j \gamma_{ij} = 0 \ .$$

They derive asymptotic tests of significance similar to an analysis of variance. It is of some interest to examine the results obtained from the application of likelihood methods to this model.

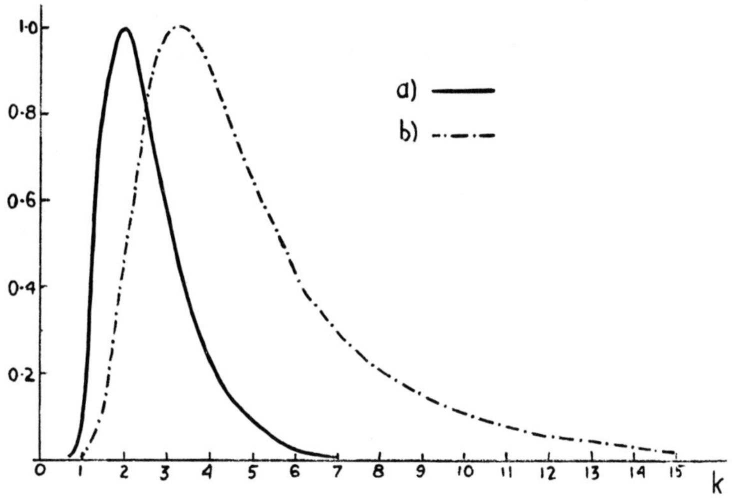

Figure 6. (a) Conditional relative likelihood of k and (b) maximum relative likelihood of k for the data of Example 3.

Restricting attention to two-parameter contagious distributions, the unconditional maximum of the likelihood can be found fairly simply when the sample mean is the maximum likelihood estimate of the population mean η. The class of such distributions seems fairly large [15].

Example 4: Negative Binomial. Suppose there is a two-way classification from negative binomial distributions defined by parameters k_{ij}, p_{ij}, $(i = 1,2,\ldots,r;\ j = 1,2,\ldots,c)$. The mean η_{ij} is then, from (4.1),

$$\eta_{ij} = k_{ij}p_{ij} = \mu+\rho_i+\beta_j+\gamma_{ij} \; .$$

The parameters k_{ij} can be examined for this example either by maximum relative likelihoods (3.4) or by conditional likelihoods (3.3). The results are essentially the same either way. Separate graphs for all six k_{ij} indicate that a common value $k_{ij} = k$ is plausible. A likelihood for the common k shows that the values of k between 0.54 and 1.08 are reasonably plausible. The likelihoods are skewed so that large sample theory would not give very good approximations.

In the calculations that follow the estimates $\hat{k}_{ij}$ obtained from the conditional likelihood are used. The likelihoods of the remaining parameters are then calculated conditionally on

$$\hat{\mu} = \bar{a},\ \hat{\rho}_i = \bar{a}_{i*} - \bar{a},\ \hat{\beta}_j = \bar{a}_{*j} - \bar{a},\ \gamma_{ij} = \bar{a}_{ij} - \bar{a}_{i*} - \bar{a}_{*j} + \bar{a}.$$

The maximum of the likelihood can therefore easily be calculated from these estimates and $\hat{k}_{ij}$.

4.1 Likelihoods of Main Effects.

In order to calculate the maximum relative likelihood function of ρ_1 it is necessary to find the restricted maximum likelihood estimates ρ_i' $(i \neq 1)$, β_j', γ_{ij}' for the fixed ρ_1, as outlined in Section 3.2 and equation (3.1). This can be done by differentiating (4.2) with respect to η_{ij} subject to the single restriction

$$f = \bar{\eta}_{1*} - \bar{\eta} - \rho_1 = 0, \tag{4.3}$$

where

$$\bar{\eta}_{1*} = \frac{1}{c}\sum_{j=1}^{c} \eta_{1j},\quad \bar{\eta} = \frac{1}{rc}\sum_{i,j} \eta_{ij}.$$

This leads to the following sets of equations:

$$\frac{\lambda}{rc}\eta_{ij}^2 + \eta_{ij}\left(\frac{\lambda}{rc}k_{ij} + n_{ij}k_{ij}\right) - T_{ij}k_{ij} = 0,\ (i \neq 1), \tag{4.4}$$

$$-\frac{r-1}{rc}\lambda\eta_{1j}^2 + \eta_{1j}\left(-\frac{r-1}{rc}\lambda k_{1j} + n_{1j}\right) - T_{1j}k_{1j} = 0,$$

where λ is the Lagrange multiplier associated with the restriction (4.3). From the above equations (4.4), $\partial\eta_{ij}/\partial\lambda$ can be calculated as

$$\frac{-\eta_{ij}^2 - k_{ij}\eta_{ij}}{2\lambda\eta_{ij} + \lambda k_{ij} + \eta_{ij}k_{ij}rc} \quad (i \neq 1)$$

and

$$\frac{-(r-1)\eta_{1j}^2 - (r-1)k_{1j}\eta_{1j}}{2(r-1)\lambda\eta_{1j} + (r-1)\lambda k_{1j} - \eta_{1j}k_{1j}\ rc} \tag{4.5}$$

From (4.3), (4.4), and (4.5),

$$\frac{\partial f}{\partial\lambda} = \frac{r-1}{rc}\sum_j \frac{\partial\eta_{1j}}{\partial\lambda} - \frac{1}{rc}\sum_{i\neq 1}\left[\sum_j\left(\frac{\partial\eta_{ij}}{\partial\lambda}\right)\right] \tag{4.6}$$

can be calculated. Given a trial value of λ, η_{ij} and $\partial\eta_{ij}/\partial\lambda$ can be calculated from (4.4) and (4.5), respectively, as functions of the estimate $\hat{k}_{ij}$ of k_{ij}; then f and $\partial f/\partial\lambda$ can be calculated from (4.3) and (4.6). The next approximation to λ is then

$$\lambda' = \lambda - f/(\partial f/\partial\lambda).$$

In this way the maximum likelihood estimates $\eta_{ij}(\rho_1)$ can be calculated for the given ρ_1. Inserting these $\eta_{ij}(\rho_1)$ and $\hat{k}_{ij}$ into (4.2) gives the maximum of the log likelihood of ρ_1. Subtracting from this the unconditional maximum of the log likelihood obtained previously and taking exponentials yields the maximum relative likelihood of ρ_1. This can then be plotted as a function of ρ_1.

Although the above formulae look complicated, the computations can be very simply programmed for a high-speed computer. This was done for the data presented in Table 2 of [10]. In their example there were $r = 2$ chemicals and $c = 3$ methods. The maximum relative likelihood of ρ_1 is plotted in Figure 7. The calculations required to generate this curve (having been given the estimates $\hat{k}_{ij}$) required less than 2 seconds on an IBM 360-75 computer. It can be seen that Figure 7 is essentially in agreement with the large sample test of significance given [10] in Table 3. However Figure 7 is somewhat more informative than merely saying chemicals (ρ_i) are significantly different from zero. The actual size of ρ_1, in fact, is given along with plausibility ratings.

The preceding formulae (and the same computer program) will produce curves for all the main-effect parameters merely by changing the order in which the observations are read into the computer. The curves for β_1 and β_2 are given in Figure 8, from which it can be seen that they also are in agreement with the conclusion obtained [10], (Table 4), i.e., no significant effect was noted due to variations in method.

Although these maximum likelihoods functions are reasonably normal in shape it seems unnecessary to calculate the large sample approximations.

4.2 Likelihoods of Interactions.

To ascertain the effect of the interactions, it is necessary to calculate the maximum relative likelihood of $\gamma_{ij} = 0$ (all i,j). There are two restrictions: $\Sigma\rho_i = 0$, $\Sigma\beta_j = 0$; but the Lagrange multipliers associated with these are zero. The equations of estimation for the maximum likelihood estimate μ'', ρ_i'', β_j'' subject to $\gamma_{ij} = 0$ are

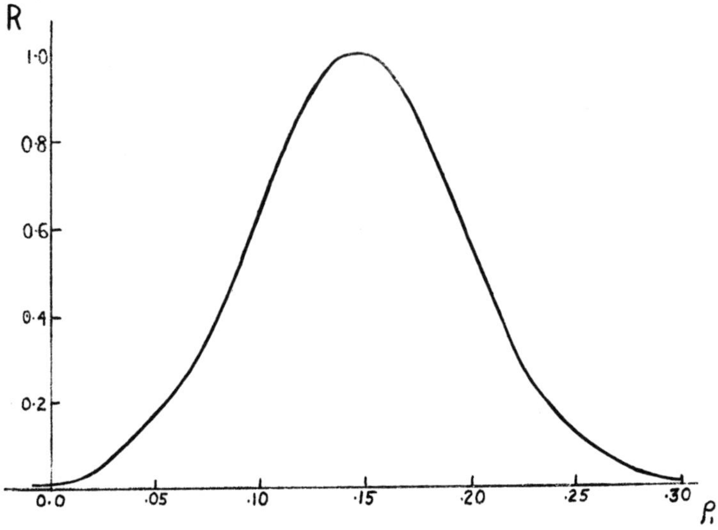

Figure 7. Maximum relative likelihood of ρ_1 for the data of Example 4.

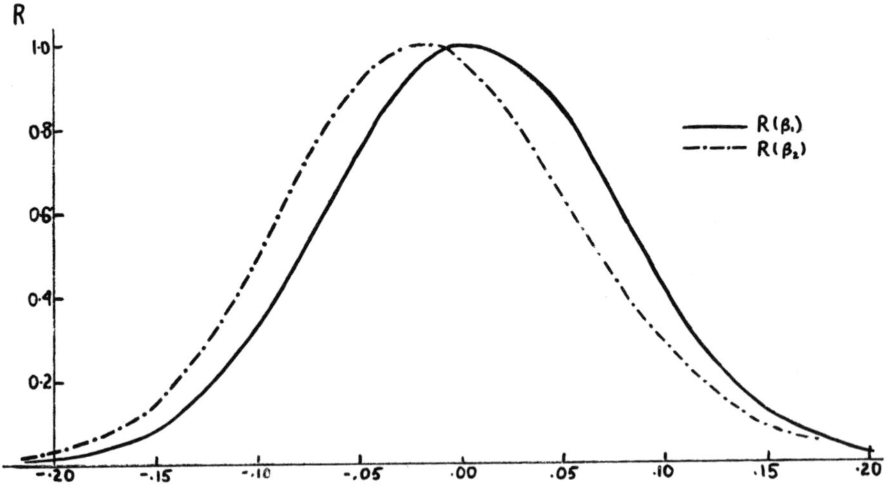

Figure 8. Maximum relative likelihoods of β_1 and β_2 for the data of Example 4.

$$L_i = \sum_j \left(\frac{T_{ij}}{\eta''_{ij}} - \frac{n_{ij}k_{ij}+T_{ij}}{k_{ij}+\eta''_{ij}} \right) = 0, \qquad i = 1,\ldots,r-1$$

$$L_{r+j-1} = \sum_i \left(\frac{T_{ij}}{\eta''_{ij}} - \frac{n_{ij}k_{ij}+T_{ij}}{k_{ij}+\eta''_{ij}} \right) = 0, \quad j = 1,\ldots,c-1$$

$$L_{r+c-1} = \Sigma\rho''_i = 0;$$

$$L_{r+c} = \Sigma\beta''_j = 0,$$

where

$$\eta''_{ij} = \mu''+\rho''_i+\beta''_j.$$

From this, the (r+c+1) x (r+c+1) matrix D of derivatives of the L's with respect to the μ'', ρ''_i and β''_j can be obtained. Given any set of trial values μ^o, ρ^o_i, β^o_j satisfying the restrictions, the L's and D can be calculated for $\hat{k}_{ij}$. A closer approximation is then given by adding the vector of corrections: $-D^{-1}L$. From the estimates μ'', ρ''_i, β''_j, $\hat{k}_{ij}$, the maximum relative likelihood of $\gamma_{ij} = 0$ can be calculated as before.

For the data in [10], twenty iterations made all but one of the L_i of the order 10^{-14}. The estimates are $\mu'' = 0.644$, $\rho''_1 = 0.160$, $\rho''_2 = -0.133$, $\beta''_1 = 0.0123$, $\beta''_2 = -0.0690$, and $\beta''_3 = 0.0567$. Thus ρ''_1 and ρ''_2 have not been very accurately determined, since $\rho''_1 + \rho''_2$ is not zero. The maximum relative likelihood of $\gamma_{ij} = 0$ is 0.0265, in agreement with Table 5 of [10]. The value $\gamma_{ij} = 0$ is relatively implausible so that the interactions do have an appreciable effect. The twenty iterations required to determine the estimates and the calculations of the maximum relative likelihood required less than 4 seconds.

The preceding likelihoods could easily be recalculated using different plausible values of the k_{ij}. This would give some idea how inferences concerning ρ_i, β_i, etc., are affected by the k_{ij}. On a high-speed computer no further programing is necessary to accomplish this.

5. EXACT TESTS OF SIGNIFICANCE

Exact goodness-of-fit tests may be obtained for certain discrete distributions by enumeration (e.g., with a computer).

Suppose that there are sufficient statistics T for all the unknown parameters θ. Let a_i be the frequency with which i is observed; the distribution of the observations is

$$\frac{n!}{\pi_i a_i!}\pi\left(p_i\right)^{a_i}(\theta),$$

where $n = \Sigma a_i$,

and $p_i(\theta)$ is the probability of observing i. If T is a sufficient statistic with distribution $p(T;\theta)$, the conditional distribution

$$p(a_0,a_1,\ldots,|T), = \frac{n!\pi_i\left(p_i\right)^{a_i}(\theta)}{\pi_i a_i!p(T;\theta)} \tag{5.1}$$

is independent of θ. Thus the probability of the observed configuration $a_0,a_1,\ldots$, as well as of any other configuration, can be calculated. If the observed configuration is highly probable, no further calculations need be made.

A number of goodness-of-fit criteria could be used. For example, one might calculate the total probability of all configurations as probable or less probable than the observed one. This usually requires too much computation even for a fast computer. A simpler criterion involves the use of some measure of departure such as

$$X^2 = \Sigma\frac{(a_i-e_i)^2}{e_i} \quad (e_i = \text{expected frequency of } i),$$

which in large samples has the χ^2 distribution. The probability of obtaining a value X^2 greater than the observed value X_0^2 can sometimes easily be exactly computed using (5.1) and enumerating the appropriate configurations. (This is discussed and illustrated in the next section.) Such tests of significance are related to the conditional likelihoods of Section 3.3; in both, the unwanted parameters have been eliminated by using sufficient statistics.

5.1. Probability distribution of X^2.

Since T is a sufficient statistic the e_i will not change in samples drawn from the conditional distribution (5.1). Also only values of X^2 less than or equal to X_0^2 need be considered, so that bounds can immediately be placed on the a_i. Since for all i

$$\frac{(a_i - e_i)^2}{e_i} \le X_0^2 ,$$

the a_i must lie between the nearest integer bounds $e_i \pm \sqrt{e_i} X_0$. This reduces the number of configurations that need to be enumerated. All $a_2, a_3 \ldots$, lying within these bounds can be systematically enumerated, and a_0, a_1 can be determined from the fixed observed values n and T. Of course a_0, a_1 must also be discarded if they do not lie in their respective intervals. When an admissible configuration has been found it can easily be ascertained if the resulting X^2 is less than or equal to X_0^2. If so, the probability of the configuration can be calculated and accumulated. The result gives the exact probability with which values X^2 will be less than or equal to the observed X_0^2.

Example 5: (Fisher, [4]) Poisson distribution:

$$p_i(\theta) = \frac{\theta^i e^{-\theta}}{i!} ,$$

$$p(T;\theta) = \frac{(n\theta)^T e^{-n\theta}}{T!} .$$

The conditional distribution (5.1) is

$$\frac{n!T!}{n^T} \left[\frac{1}{\pi_i (a_i)! \ \pi_i (i!) a_i} \right].$$

The observations were $a_0 = 124$, $a_1 = 12$, $a_2 = 2$, $a_3 = 2$, giving $n = 140$ and $T = 22$. Then $e_0 = 119.6415$, $e_1 = 18.8008$, $e_2 = 1.4772$, $e_3 = 0.0805$, and $X_0^2 = 48.564$. Upper limits for a_2, a_3 are 9 and 2. The probability that $X^2 < X_0^2$ is 0.9973, giving a significance level of 0.00268. The calculations required 3 seconds.

Two larger examples were tried, as well. The distribution $a_0 = 80$, $a_1 = 61$, $a_2 = 31$, $a_3 = 1$, $a_4 = 0$ ([14] p. 226) required 50 seconds; and the distribution $a_0 = 8$, $a_1 = 16$, $a_2 = 18$, $a_3 = 15$, $a_4 = 9$, $a_5 = 7$ ([2] p. 123) required less than 6 seconds.

Example 6: One of the negative binomial distributions ($k = 2$) generated for Example 3 had the following frequencies $a_0 = 2$, $a_2 = 1$, $a_5 = 1$, $a_7 = 1$, and all other a's were zero. Thus $n = 5$, $T = 14$. The conditional distribution (5.1) is

$$\frac{n!\prod_i \binom{k+i-1}{k-1}^{a_i}}{\prod_i a_i! \binom{nk+T-1}{nk-1}} = \frac{T!n! \left[k^{n-a_0}(k+1)^{n-a_0-a_1}\ldots\right] (nk-1)!}{\prod a_i! \prod (i!)^{a_i} (nk+T-1)!} .$$

The observed X_0^2 is 9.884, so that upper bounds for possible configurations are 3,3,2,2,1,1,1 for $a_1, a_2, \ldots, a_7$. This leads to a probability of $(X^2 < X_0^2) = 0.6619$, so that the significance level of the observations is 34%. The computations took about 4 seconds. Another example of similar size also required only 4 seconds. Of course not all samples will yield such fast results; and some will be prohibitively long, but usually this will happen only in large samples where more conventional methods work.

Example 7: A particularly simple example occurs in genetics. The observable genotypes MM, MN, NN have probabilities p^2, $2pq$, q^2, respectively, under the hypothesis of random mating, where p is the frequency of the M gene and $q = 1-p$ is the frequency of the N gene. If the observed frequencies are a_1, a_2 and a_3, then the probability distribution is

$$p(a_1,a_2,a_3) = \frac{n!}{a_1!a_2!a_3!} p^{2a_1} (2pq)^{a_2} (q)^{2a_3} ,$$

where $a_1+a_2+a_3 = n$. Thus $T = 2a_1+a_2$ is a sufficient statistic for p. The distribution of T is

$$\frac{(2n)!}{T!(2n-T)!} \left(p^T q^{2n-T}\right) .$$

The conditional distribution of a_1, a_2, a_3 given T (5.1), is

$$p(a_1,a_2,a_3| T) = \frac{T!(2n-T)!n!2^{a_2}}{(2n)!a_1!a_2!a_3!} .$$

Thus a goodness-of-fit test can be computed in any particular case as described in Sections 5 and 5.1, examples 5 and 6. This is particularly useful when q is small, as in the case of a rare trait. Then a_3 and its expected value are small so that the usual χ^2 goodness-of-fit test cannot be performed (e.g. [13], pp. 13-14, 22-25.). It can be seen that

$$a_2 = T-2a_1$$

$$a_3 = n+a_1-T,$$

so that

$$p(a_1,a_2,a_3|T) = \frac{T!(2n-T)!n!2^{T-2a_1}}{(2n)!a_1!(T-2a_1)!(n+a_1-T)!} ; \qquad (5.2)$$

also

$$\max(0,T-n) \le a_1 \le T/2.$$

Thus the enumeration in this case is simple to perform even on a desk calculator.

The following frequencies of the M and N blood groups were taken from [13], p. 25: $a_1 = 475$, $a_2 = 89$, $a_3 = 5$. Thus $T = 1039$, $n = 569$, $T-n = 490$ and $T/2 = 519.5$, so that the range of a_1 is $490 \le a_1 \le 519$. Here the range of a_1 is so small that the entire conditional distribution (5.2) can be tabulated. The probability of the observed configuration and less probable ones can be calculated giving a significance level of 60.3%. These calculations required less than three seconds.

6. SIMULATED TESTS OF SIGNIFICANCE.

Usually it is not possible to perform exact tests as in Section 5 because there are no sufficient statistics, or the enumeration is too long. It may still be possible, however, to put bounds on the significance level by simulating it on a computer.

Suppose the probability that an observation falls in class i is $p_i(\theta)$ and the observed frequency is a_i $(i = 1,2,...)$. Suppose

$$\Sigma a_i = n$$

and

$$P_i(\theta) = \sum_{j=1}^{i} p_i(\theta), \quad (i = 1,2,...), \quad P_0(\theta) = 0.$$

Suppose $\hat{\theta}$ is the maximum likelihood estimate of θ. The observed value X_0^2 of the usual χ^2 statistic can be calculated

$$X_o^2 = \sum(a_i-e_i)^2/e_i,$$

where $e_i = np_i(\hat{\theta})$.

The probability of obtaining $X^2(\theta) \ge X_o^2$ for any specified θ can be estimated by simulation on a computer as follows. A random sample of n numbers is selected from the uniform distribution

(0,1). Suppose for the specified θ, o_i of these numbers lie between $P_{i-1}(\theta)$ and $P_i(\theta)$ $(i=1,2,\ldots)$. The o_i are then a random sample of size n from the multinomial distribution defined by the $p_i(\theta)$. Given these o_i, a new maximum likelihood estimate θ' can be calculated and, from this, new values $e_i' = np_i(\theta')$ obtained. These o_i and e_i' determine a new value of $X^2(\theta)$ (which of course depends on the specific value θ used to generate the $P_i(\theta)$):

$$X^2(\theta) = \sum(o_i-e_i')^2/e_i'.$$

This procedure can be repeated a large number of times on a computer using the same value of θ and hence the frequency with which the $X^2(\theta)$ exceed the observed X_o^2 obtained. In this way an estimate of the significance level of X_o^2 can be obtained for the given θ. Similarly, estimates of the significance level for various values of θ can be calculated, and, in turn, the upper bound for the significance level can be found.

It should be noted that, if n is sufficiently large, the $X^2(\theta)$ has approximately the χ^2 distribution so that $\Pr(X^2(\theta) \geq X_o^2)$ becomes approximately independent of θ.

It should also be noted that the empirical test of this section does not yield significance levels of X^2 in the <u>absence of knowledge of θ</u>. Rather, it is a procedure for examining empirically the significance levels of X^2 for <u>all specified θ</u> (or at least for all plausible values of θ). Appropriate tests of significance for X^2 (or allied criteria) valid in the absence of knowledge of θ require the existence of sufficient estimation of θ and are given in Section 5.

<u>Example 8: Negative Binomial</u>: If k is unknown the exact test of Section 5 cannot be applied. However, the maximum likelihood estimate $\hat{k}$ is the solution of a polynomial equation and can be found very quickly using iteration. This was applied to the data in Table V, [9]. The maximum likelihood estimate of k was $\hat{k} = 5.2479$, and the observed X_0^2 was 4.857 (with no grouping). Seventy trials of the above procedure for a given θ were obtained in 10 seconds of computer time, of which 35 trials yielded values of X^2 greater than the observed X_0^2. A second run of 10 seconds produced 39 greater out of a total of 69 trials. This gives an estimated significance level of 0.53, and it is extremely unlikely that the exact significance level is outside of the range (0.40, 0.65). Other values of θ produced essentially the same results.

This is in accordance with the results of [9].

The goodness-of-fit of the observations in [10] to a negative binomial distribution was checked this way. It was found that about 75 trials could be performed in 10 seconds in each case. Variation of θ produced little difference in the results. The computations in Sections 5 and 6 were done on an IBM 360-75 computer.

ACKNOWLEDGEMENT

I should like to thank Professor G. A. Barnard, John D. Kalbfleisch, James G. Kalbfleisch, and V. P. Godambe for reading the manuscript and for their criticisms and suggestions.

REFERENCES

[1] Barnard, G. A.; Jenkins, G. M.; and Winston, C. B. 1962. Likelihood inference and time series. J. R. Statist. Soc. B 125:321-372.

[2] Feller, W. 1950. An Introduction to Probability Theory and Its Applications. 1st edition, 123, John Wiley and Sons, New York.

[3] Fisher, R. A. 1935. The logic of inductive inference. J. R. Statist. Soc. 98:39-54.

[4] ————. 1950. The significance of deviations from expectation in a Poisson series. Biometrics 6:17-24.

[5] ————. 1956. Statistical Methods and Scientific Inference. Gordon, Oliver and Boyd.

[6] ————., and Yates, F. 1963. Statistical Tables for Biological Agricultural and Medical Research. Hafner Publishing Co., N.Y., 6th edition.

[7] Fraser, D. A. S., 1967. Data transformations and the linear model. Ann. Math. Statist. 38:1456-1465.

[8] ————. 1968. The Structure of Inference. John Wiley and Sons, New York.

[9] Gurland, J., and Hinz, P. 1968. Testing fit and analysing untransformed data for the negative binomial distribution. The Future of Statistics. (ed. D. G. Watts) Academic Press, New York and London, 163-174.

[10] Hinz, P., and Gurland, J. 1968. A method of analysing untransformed data from the negative binomial and other contagious distributions. Biometrika 55:163-170.

[11] Kalbfleisch, J. D. and Sprott, D. A. 1969. Application of likelihood methods to models involving large numbers of parameters. (submitted).

[12] Katti, S. K., and Gurland, J. 1961. The Poisson-Pascal distribution. Biometrics 17:527-538.

[13] Li, C. C. 1954. Population Genetics. The University of Chicago Press.

[14] Snedecor, G. W. and Cochran, W. G. 1967. Statistical Methods. 6th edition, the Iowa State University Press, Ames, Iowa.

[15] Sprott, D. A. 1965. A class of contagious distributions and maximum likelihood estimation. Proc. classical and contagious discrete distributions, 337-350, Pergamon Press, London.

[16] ——————. , and Kalbfleisch, J. G. 1965. Use of the likelihood function in inference. Psychol. Bull. 64, 15-22.

[17] ——————., and Kalbfleisch, John B. 1969a. Examples of likelihoods and comparison with point estimates and large sample approximations. To be published by Jour. Amer. Statist. Ass.

[18] ——————., and ——————. 1969b. Marginal and Conditional Likelihood. (Submitted for publication).

METHODS FOR ELIMINATING ZERO COUNTS IN CONTINGENCY TABLES

STEPHEN E. FIENBERG
The University of Chicago

PAUL W. HOLLAND*
Harvard University

SUMMARY

In this selection a class of estimators of the cell expectations of contingency tables is proposed as a general-purpose solution to the problem of adjusting zero counts in observed cross-classifications. On the same set of data the effects of the several estimators are illustrated, and the large-sample behavior of these estimators is derived. Some exact small-sample results are given for one estimator. Finally, possible extensions and improvements are suggested.

* Work facilitated by National Science Foundation grant (GS-2044X).

1. INTRODUCTION

In the analysis of counted data it is not uncommon to encounter problematic zeros. For example, when one is transforming counts to the various linearizing scales such as logarithms, logits or probits, some "adjustment" must be made whenever a zero count is encountered.

Zero counts may also be troublesome because, paradoxically, some zeros are smaller than others. For example, if 0/5 and 0/500 are two observed death rates, it may be a misleading practice to report them as equal since they carry quite different information. Another version of this same problem arises with the arc-sine transformation for stabilizing the variance of proportions. The simple arc-sine transform will assign zero to a $\hat{p}$ of zero no matter if n = 5 or 500, and yet 0/5 and 0/500 do not look very equal when binomial variation is considered.

For some of these examples there are devices that eliminate the problem of zeros. [14] Summarizes various attempts to improve the sampling properties of estimators of the logit transform used in bioassay. All of the estimators of log(p/q) which they discuss make adjustments that avoid infinity whenever $\hat{p}$ or $\hat{q}$ is zero. For the arc-sine transform, the Freeman-Tukey [12] adjustment, which stabilizes the variance of proportions even in relatively small samples, also has the property that the value assigned to a zero proportion depends on the value of n--and the larger the n, the smaller the value assigned to zero.

In these two examples, the treatment of zero counts arises as a by-product of other aims. We shall take the opposite approach and shall suggest several methods for eliminating zero counts in cross-tabulations which can be used prior to other types of analyses. Because we want a "general-purpose" solution to this problem we choose to view it as the simultaneous estimation of all the cell expectations. Although it is possible to apply our results to other situations, we shall only consider counted data in the form of a cross-tabulation--hereafter simply referred to as a <u>table</u>. For purposes of unity and simplicity, we shall couch our discussion in terms of a 2-way, r by c, table. We note, however, that sometimes what we say is easily extended to higher dimensional tables, or may be more <u>naturally</u> discussed with reference to a simple multinomial variable with no cross-classification structure at all.

Before proceeding, it is useful to distinguish between two types of zero entries in observed tables. We shall call these <u>fixed</u> and <u>random</u> zeros. Fixed zeros occur in cells like "Male Obstetrical Patients": the zero is due to an honest zero probability. Random zeros occur in sparce but non-empty categories like "Jewish Farmers from Iowa". The main point is that an estimation procedure should leave fixed zeros alone, but might change a random zero to something positive. Other writers have been concerned with estimation in the presence of fixed zeros (see [17], [19] and [6]); and while we are mostly interested in random zeros since these are the ones that ought to be adjusted, we shall indicate briefly how our estimators can allow for both types simultaneously.

2. SMOOTHING TABLES WITH PSEUDO-COUNTS

Researchers in the various substantive disciplines are often reluctant to "adjust" their raw data because it suggests manipulation of the "facts". Statisticians have fewer qualms in this regard. Thus, one method for eliminating zeros in counted data that appears in various guises throughout the statistical literature is what we shall call "smoothing with pseudo-counts". To illustrate this technique, consider the aritificial cross-tabulation given in Table 1.

TABLE 1

An artificial example of a cross-classification with two empty cells.

0	3	5	8
4	15	25	44
4	26	0	30
8	44	30	82

To rid Table 1 of its zeros we add 6 fake counts to the total of 82 and distribute these 6 equally to the 9 cells. This adds 2/3 of a pseudo-count to each cell. The result is given in Table 2a.

To preserve the original total of 82 we multiply each entry of Table 2a by 82/88. This results in the final "smoothed" table shown in Table 2b.

TABLE 2

(a) Table 1 after the addition of 0.67 pseudo-counts to each cell.

(b) Final result of shrinking Table 1 towards the constant table.

(a)

0.67	3.67	5.67	10.01
4.67	15.67	25.67	46.01
4.67	26.67	0.67	32.01
10.01	46.01	32.01	88.03

(b)

0.6	3.4	5.3	9.3
4.4	14.6	23.9	42.9
4.4	24.8	0.6	29.8
9.4	42.8	29.8	82.0

There are two mysterious items in going from Table 1 to Table 2b. First, why 6 fake counts? Second, why were the 6 pseudo-counts distributed equally among the nine cells? Also note that this use of pseudo-counts only preserved the grand total, 82. Not only have we altered every entry of Table 1, but we have also altered each of the marginal totals as well. On the other hand, the overall change from Table 1 to Table 2b is small.

The basic problem one encounters in applying the method of pseudo-counts lies in choosing the number to add to each cell--in general, different numbers could be added to different cells. In practice, we fear, the number of pseudo-counts one adds to his tables is part of the "research folklore" of his discipline; like some other statistical methods, the criteria involved in such manipulations are shrouded in mystery.

[15] Gives an excellent historical account of the method of pseudo-counts, [22] generalizes Laplace's Law of Succession and added one pseudo-count to each cell; [1] suggests adding 1/4 of a pseudo-count but only to the empty cells; [8] indicates that such a procedure gave reasonable results; more recently, [20] and [13] have proposed 1/2 of a pseudo-count added to all cells, while [27] suggests adding 1/2 count to the empty cells only. The introductory remarks in [14] give further references for these choices in the logit case.

Although 1, 1/2 and 1/4 are popular numbers, there is a general lack of agreement in the literature as to which one to use and whether to add it to every cell or only to those that are empty. After some light theoretical development, we shall suggest ways of using pseudo-counts that depend on the observed table.

3. PSEUDO-COUNTS AND BAYES ESTIMATORS

Denote the observed table by $\underset{\sim}{X}$ and its $(i,j)^{th}$ entry by X_{ij}, $i = 1,\ldots,r$, and $j = 1,\ldots,c$. For simplicity, we assume $\underset{\sim}{X}$ has a multinomial distribution with cell probability matrix, $\underset{\sim}{p}$, and sample size $N = \sum_{i,j} X_{ij}$. Therefore

$$P\{\underset{\sim}{X}=\underset{\sim}{x}|\underset{\sim}{p},N\} = N! \prod_{i,j}\left[\frac{p_{ij}^{x_{ij}}}{x_{ij}!}\right], \tag{3.1}$$

where $p_{ij} > 0$ and $\sum_{i,j} p_{ij} = 1$.

Suppose $\underset{\sim}{T} = \underset{\sim}{T}(\underset{\sim}{X})$ is a "smoothed version" of the given table, $\underset{\sim}{X}$. $\underset{\sim}{T}$ may be thought of as an estimator of $N\underset{\sim}{p}$, which is the expectation of $\underset{\sim}{X}$, given $\underset{\sim}{p}$. As our loss function, we adopt the squared Euclidean distance from $\underset{\sim}{T}(\underset{\sim}{X})$ to $N\underset{\sim}{p}$, i.e.,

$$\|\underset{\sim}{T}(\underset{\sim}{X})-N\underset{\sim}{p}\|^2 = \sum_{i,j}(T_{ij}(\underset{\sim}{X})-Np_{ij})^2. \tag{3.2}$$

The risk function of $\underset{\sim}{T}$ at $\underset{\sim}{p}$ is given by

$$R(\underset{\sim}{p};\underset{\sim}{T}) = E_p\|\underset{\sim}{T}(\underset{\sim}{X})-N\underset{\sim}{p}\|^2. \tag{3.3}$$

The expectation is taken over the conditional distribution of $\underset{\sim}{X}$, given $\underset{\sim}{p}$. $R(\underset{\sim}{p};\underset{\sim}{T})$ is the average squared distance of $\underset{\sim}{T}$ from $N\underset{\sim}{p}$; thus the smaller the risk, the better the estimator. For the raw table, $\underset{\sim}{X}$, it is easy to show that the risk function is

$$R(\underset{\sim}{p};\underset{\sim}{X}) = N[1-\|\underset{\sim}{p}\|^2]. \tag{3.4}$$

Now suppose we give $\underset{\sim}{p}$ a prior density, $\pi(\underset{\sim}{p})$. The <u>Bayes risk</u> of an estimator, $\underset{\sim}{T}$, with respect to π, is

$$E_\pi[R(\underset{\sim}{p};\underset{\sim}{T})]. \tag{3.5}$$

If an estimator, $\underset{\sim}{B}$, minimizes (3.5) among all possible estimators of $N\underset{\sim}{p}$, then $\underset{\sim}{B}$ is <u>Bayes against</u> π. For any estimator, $\underset{\sim}{T}$, $R(\underset{\sim}{p};\underset{\sim}{T})$ is a polynomial in the p_{ij} and is therefore continuous. Thus, all Bayes estimators are admissible for any prior density which is absolutely continuous (see [2]). For such estimators no other estimator of $N\underset{\sim}{p}$ can have a uniformly smaller risk function for all $\underset{\sim}{p}$. With the quadratic loss function (3.2), it well known that $\underset{\sim}{B}$ is Bayes against π if

$$B_{ij}(\underset{\sim}{X}) = E(Np_{ij}|\underset{\sim}{X}). \tag{3.6}$$

That is, the Bayes estimators are the posterior regressions of Np_{ij} on $\underset{\sim}{X}$.

Let $\lambda_{ij} = E_{\pi}(p_{ij})$ and $\underset{\sim}{\lambda} = (\lambda_{ij})$. Then we have

$$E(X_{ij}) = N\lambda_{ij}, \tag{3.7}$$

so that $N\underset{\sim}{\lambda}$ is the unconditional expectation of both $\underset{\sim}{X}$ and $N\underset{\sim}{p}$, and $\sum_{i,j} \lambda_{ij}$ is equal to unity.

If $\pi(\underset{\sim}{p})$ is proportional to

$$\prod_{i,j} p_{ij}^{k\lambda_{ij}-1} \tag{3.8}$$

for some positive k, then p has a Dirichlet prior. It is well known (see [15]) that for this choice of π, $\underset{\sim}{B}$ has the form:

$$\underset{\sim}{B}(\underset{\sim}{X}) = \frac{N}{N+k}(\underset{\sim}{X}+k\underset{\sim}{\lambda}). \tag{3.9}$$

However, (3.9) may be interpreted as the method of pseudo-counts where a total of k counts is added to the table, with $k\lambda_{ij}$ going into the $(i,j)^{th}$ cell, and the whole table is then renormalized to preserve the grand total of N. When k and $\underset{\sim}{\lambda}$ are fixed and do not depend on $\underset{\sim}{X}$, the above shows that the method of pseudo-counts leads to estimators of the cell expectations that are admissible with respect to quadratic loss. Furthermore, from this point of view, the difficulties in choosing the number of pseudo-counts to add to each cell are seen to be the same difficulties that arise in the choice of prior distributions. To avoid some of these problems, we propose to make k and $\underset{\sim}{\lambda}$ depend on $\underset{\sim}{X}$ in "reasonable" ways. Of course, if k and $\underset{\sim}{\lambda}$ are functions of X, we cannot apply the admissibility facts mentioned above to the resulting estimators. On the other hand, we shall show that such estimators have reasonably good sampling properties and therefore are useful solutions to the problem of eliminating zero counts. In the next section, we shall propose a way of choosing k that depends on $\underset{\sim}{X}$ and $\underset{\sim}{\lambda}$, and we shall suggest various choices of λ, some of which also depend on X.

We end the present section with a technical observation. One might suppose that the method of pseudo-counts is a Bayes solution for more types of priors than the Dirichlet. After all, (3.9) is just a special kind of linear regression of p on X. Although this may be true, the following theorem indicates that it is much less general than might be conjectured.

Theorem 1: Suppose that: (a) X has a multinomial distribution with parameters N and p; (b) p has a prior density $\pi(p)$, not depending on N, with $\lambda = E_{\pi}(p)$; and (c) there exists a k, independent of N, such that for $i = 1,2,\ldots,r$ and $j = 1,2,\ldots,c$,

$$E(p_{ij}|X) = \frac{X_{ij}+k\lambda_{ij}}{N+k} .$$

Then $\pi(p)$ is the Dirichlet distribution.

The proof of Theorem 1 consists of showing that all the cross-moments of p are determined by λ and k. The essential ingredients of the argument can be found in [16], page 404.

4. ON CHOOSING k AND λ

If we evaluate the risk function (3.3) for the Bayes estimator (3.9) that corresponds to putting $k\lambda_{ij}$ pseudo-counts into cell (i,j), we obtain

$$R(p;B) = \left(\frac{N}{N+k}\right)^2 [N(1-\|p\|^2) + k^2\|\lambda-p\|^2]. \tag{4.1}$$

For fixed p and λ the value of k that minimizes (4.1) is given by

$$k^*(p;\lambda) = \frac{(1-\|p\|^2)}{\|\lambda-p\|^2} . \tag{4.2}$$

The value of R(p;B), when $k = k^*$, is

$$\frac{N}{N+k^*}[N(1-\|p\|^2)].$$

The expression in brackets is exactly the risk function of the raw, unsmoothed table, X. Thus, if we could choose k optimally, the result would always improve on the risk of X. Unfortunately, k^* depends on p, which is unknown, so this improvement is not actually realizable. What we propose to do is to substitute $\|\frac{1}{N}\underset{\sim}{X}\|$ for $\|\underset{\sim}{p}\|$ and $\|\underset{\sim}{\lambda}-\frac{1}{N}\underset{\sim}{X}\|$ for $\|\underset{\sim}{\lambda}-\underset{\sim}{p}\|$ in (4.2). This defines a value of k that depends on $\underset{\sim}{X}$ and the choice of $\underset{\sim}{\lambda}$. We denote this by $\hat{k}(\underset{\sim}{\lambda})$. It is a sample estimate of the optimal value of k, so that

$$\hat{k}(\underset{\sim}{\lambda}) = \frac{N^2-\|\underset{\sim}{X}\|^2}{\|N\underset{\sim}{\lambda}-\underset{\sim}{X}\|^2} = k^*\left(\frac{1}{N}\underset{\sim}{X},\underset{\sim}{\lambda}\right). \tag{4.3}$$

The resulting estimator of $N\underset{\sim}{p}$ is

$$\underset{\sim}{B}(\underset{\sim}{\lambda};\underset{\sim}{X}) = \frac{N}{N+\hat{k}(\underset{\sim}{\lambda})}(X) + \frac{\hat{k}(\underset{\sim}{\lambda})}{N+\hat{k}(\underset{\sim}{\lambda})}(N\underset{\sim}{\lambda}). \tag{4.4}$$

$\underset{\sim}{B}(\underset{\sim}{\lambda};\underset{\sim}{X})$ is a convex combination of $\underset{\sim}{X}$ and the table, $N\underset{\sim}{\lambda}$, which shrinks $\underset{\sim}{X}$ toward $N\underset{\sim}{\lambda}$. The convexifying weight, $N/[N+\hat{k}(\underset{\sim}{\lambda})]$, is a sample estimate of the optimal weight given by $k^*(\underset{\sim}{p};\underset{\sim}{\lambda})$. For the remainder of this section we assume k is chosen as $\hat{k}(\underset{\sim}{\lambda})$ so that we are left with the problem of choosing $\underset{\sim}{\lambda}$.

Let us return to our origninal example of smoothing Table 1 to obtain Table 2b. According to our present notation we chose $k = 6$ and the λ_{ij} all equal to 1/9. If we were to use our value of k, $\hat{k}(\underset{\sim}{\lambda})$, what should we have done with Table 1? When $\lambda_{ij} = 1/rc$, $\hat{k}(\underset{\sim}{\lambda})$ simplifies to

$$k_1 = \frac{N^2-\|X\|^2}{\|X\|^2-N^2/rc}\ . \tag{4.5}$$

For Table 1 this equals 6.07 which is accidentally very close to our original selection of the number 6. Consequently if we use $\lambda_{ij} = 1/9$ and $k = \hat{k}$, Table 1 will be smoothed to something very close to Table 2b. We shall refer to the case when the λ_{ij} are all equal as "shrinking $\underset{\sim}{X}$ toward the <u>constant table</u>".

As mentioned earlier, Table 2b does not preserve the margins of Table 1. In some cases this might be suitable, but it is not hard to think of situations where margin preserving would be desirable--for example, the margins may be fixed by experimental conditions. In order to preserve the marginal totals it is necessary that $\underset{\sim}{\lambda}$ depend on $\underset{\sim}{X}$. In particular, $N\underset{\sim}{\lambda}$ must have the same margins as $\underset{\sim}{X}$ does; i.e.,

$$\begin{aligned} \lambda_{i+} &= X_{i+}/N \qquad \text{for } i = 1,2,\ldots,r, \\ \lambda_{+j} &= X_{+j}/N \qquad \text{for } j = 1,2,\ldots,c. \end{aligned} \tag{4.6}$$

There is an infinity of values of $\tilde{\lambda}$ which satisfy these linear constraints. In particular, all values of $\tilde{\lambda}$ which satisfy (4.6) lie in the intersection of a linear manifold of dimension (r-1)(c-1) with the (rc-1) dimensional probability simplex, $S_{rc} = \{(u_{11},\ldots,u_{rc}):u_{ij} \geq 0 \text{ and } \sum_{i,j} u_{ij} = 1\}$. Each $\tilde{\lambda}$ in this intersection is determined by a unique set of values of the cross-product ratios

$$\alpha_{ij} = \frac{\lambda_{i,j}\lambda_{i+1,j+1}}{\lambda_{i,j+1}\lambda_{i+1,j}} \qquad \begin{array}{l} i = 1,\ldots,r-1, \\ j = 1,\ldots,c-1. \end{array} \tag{4.7}$$

The system of cross-ratios has been used by various authors to define the "interaction structure" of the matrix $\tilde{\lambda}$ as distinct from its margins (see [10], [18], and [23]). Choosing the α_{ij} all equal to unity corresponds to a $\tilde{\lambda}$ exhibiting perfect independence of its rows and columns. In that case, if $N\tilde{\lambda}$ also has the same marginal totals as $\tilde{X}$, for all i and j,

$$N\lambda_{ij} = X_{i+}X_{+j}/N. \tag{4.8}$$

This is the usual "expected value" for computing chi-square. For this choice of $\tilde{\lambda}$ our estimator $B(\tilde{\lambda};\tilde{X})$ amounts to shrinking the observed table $\tilde{X}$ towards its "expected value" under independence. We shall refer to this situation as "shrinking $\tilde{X}$ towards its independent projection". To illustrate this we continue our example of smoothing Table 1. Table 3a displays the "expected value" of Table 1 from the "product-of-the-margins-divided-by-N" rule.

Using Table 3a as $N\tilde{\lambda}$, $\hat{k}(\tilde{\lambda})$ may be computed from expression (4.3). It turns out to equal 13.53. This is larger than the previous value of $\hat{k}$, 6.07, because $\tilde{X}$ is closer to its independent projection than it is to the constant table with total N. If the entries of Table 3a are all multiplied by the ratio $\hat{k}(\tilde{\lambda})/N$, then we obtain the allocation of the 13.53 pseudo-counts to the individual cells. This allocation appears in Table 3b. Finally, if Table 3b is added cell by cell to Table 1 and the result multiplied by 82/(82+13.53) to preserve the grand total of 82, we obtain Table 3c.

It may be objected that both of the previous choices are obviously wrong in view of Table 1 itself. It is obviously neither constant nor independent, so why shrink $\tilde{X}$ towards either of these tables? A natural suggestion is to choose a

TABLE 3

(a) "Expected value" of Table 1 from "product of margins divided by N" rule.

(b) Table 3a normalized to have grand total of 13.53.

(c) The final result of shrinking Table 1 towards its independent projection.

(a)

0.78	4.29	2.93	8.00
4.29	23.61	16.10	44.00
2.93	16.10	10.97	30.00
8.00	44.00	30.00	82.00

(b)

0.13	0.71	0.48	1.32
0.71	3.89	2.66	7.26
0.48	2.66	1.81	4.95
1.32	7.26	4.95	13.53

(c)

0.11	3.19	4.70	8.00
4.04	16.21	23.75	44.00
3.85	24.60	1.55	30.00
8.00	44.00	30.00	82.00

table toward which to shrink $\underset{\sim}{X}$ that has an interaction structure more like that of $\underset{\sim}{X}$ itself.

The social mobility tables for Denmark and Britain, discussed in [19] and [26], are examples of tables which clearly do not exhibit independence. If one wished to eliminate a zero in a new social mobility table with the same categories as the tables discussed therein, he could choose the α_{ij}'s of $N\underset{\sim}{\lambda}$ to be more in line with the cross-product ratios of these tables.

Returning to Table 1, suppose we have looked at similar tables before, and we now decide to use the cross-product ratios from the previous data as our interaction structure for λ. Thus suppose we take $\alpha_{11} = .5, \quad \alpha_{12} = 1, \quad \alpha_{21} = 1, \quad \alpha_{22} = .05.$ (4.9)

Admittedly, these particular values of α_{ij} have "come out of the hat". They are actually rough, "eyeball" approximations to the interaction structure exhibited by the original Table 1, and the two smoothed tables, 2b and 3c. Our purpose in using these values of α_{ij} is merely to illustrate the use of a given interaction structure to determine $N\underset{\sim}{\lambda}$.

Now we need to construct $N\underset{\sim}{\lambda}$ so that it has (a) the interaction structure of (4.9) and (b) the same marginal totals as $\underset{\sim}{X}$ in Table 1. We first need to construct a table satisfying (4.9). There are many of these, and we give one in Table 4a.

TABLE 4

(a) Example of a table with the interaction structure indicated by (4.9).

(b) Final result of shrinking Table 1 towards a table with the interaction indicated by (4.9).

(a)

0.5	1	1	2.35
1	1	1	3
1	1	0.05	2.05
2.5	3	2.05	7.55

(b)

0.25	3.16	4.59	8.00
3.35	16.61	24.04	44.00
4.40	24.23	1.37	30.00
8.00	44.00	30.00	82.00

Next, we give Table 4a the margins of Table 1 via an iterative scaling procedure proposed in [9]. This scheme proceeds as follows. First, adjust the row margins of Table 4a to agree with those of Table 1 by multiplying each row of Table 4a by the ratio of the corresponding total in Table 1 to its own row total. That is, multiply each element of the first row of Table 4a by 8/2.5, the second row by 44/3 and the third row by 30/2.05. The column margins of the resulting table will not agree with those of Table 1; thus the entire procedure is repeated on the columns. After this, however, the column totals will be correct but the row totals wrong; therefore we must repeat the procedure iteratively, first for rows and then for columns. [21] And [11] have shown that this iterative scheme does converge. [4] And [26] summarize the increasing literature on the procedure found in [9]. The table we obtain after convergence has the margins of Table 1, and because the cross-product ratios are invariant under row and column multiplication, the α_{ij} are the same as in Table 4a.

Once we have computed this table, it becomes our $N\underset{\sim}{\lambda}$; using formula (4.3) we compute our value of $\hat{K}(\underset{\sim}{\lambda})$. It is 365.6 Such a large number of pseudo-counts when compared with the total number of real counts, 82, seems shocking, but it merely means that $N\tilde{\lambda}$ is close to $\tilde{X}$, so the overall change from $\tilde{X}$ to $\tilde{B}(\tilde{\lambda};\tilde{X})$ is small. We see this by inspecting Table 4b, which is the corresponding smoothed version of Table 1.

To summarize, in this section we have suggested three methods for removing random zeros in a cross-classification. The simplest is to shrink toward the constant table, that is, to add the same number of pseudo-counts to each cell. The number to be added to each cell can be computed from the observed table via formula (4.5). This procedure will usually have the smallest value of $\hat{k}$, since most real tables are not close to the constant table. This approach will not preserve marginal totals. The second procedure is to shrink $\tilde{X}$ towards its independent projection, which will preserve marginal totals; and the number of pseudo-counts added will vary from cell to cell. Formula (4.3) can be used to choose the total number of pseudo-counts to add. Both of these methods are "general-purpose" and do not require an extensive analysis of the table for their application. For many purposes they will suffice. The third method is more complicated and consists of three steps:

(a) Use of prior information to produce an approximation to the interaction structure underlying $\tilde{X}$.

(b) Iteration of a table to have this given interaction structure, as well as the margins of $\tilde{X}$. Use the result as $N\tilde{\lambda}$.

(c) Use of (4.3) to compute $\hat{k}(\tilde{\lambda})$ and (4.4) as the final estimate of $N\tilde{p}$.

5. ONLY THE EMPTY CELLS?

As indicated by our discussion of the literature, there is some question as to whether pseudo-counts should be added to all cells or only the empty ones. In the present context, adding counts only to the empty cells corresponds to shrinking $\tilde{X}$ towards an $N\tilde{\lambda}$ which has the property that $N\lambda_{ij} = 0$, unless $X_{ij} = 0$. If we add the same number to each empty cell the non-zero $N\lambda_{ij}$ are all equal and their sum is N. Let e denote the number of empty cells in $\tilde{X}$. We assume $e > 0$. For this choice of $N\tilde{\lambda}$, $\hat{k}(\tilde{\lambda})$ may be written as

$$\hat{k}_0 = \frac{N^2 - \Sigma X_{ij}^2}{\Sigma X_{ij}^2 + N^2/e} . \tag{5.1}$$

If we apply (5.1) to Table 1, we find that $\hat{k}_0 = 1.03$ so that we put about 1/2 a pseudo-count in the empty cells and none elsewhere. Table 5 gives the resulting smoothed version of Table 1.

TABLE 5

Final result of adding 0.50 pseudo-counts to the empty cells of Table 1 and renormalizing to preserve the grand total

0.51	2.96	4.94	8.41
3.95	14.81	24.69	43.45
3.95	25.68	0.51	30.14
8.41	43.45	30.14	82.00

Like smoothing $\underset{\sim}{X}$ towards the constant table, this method does not preserve the marginal totals, but the adjustment is neither large nor is it unreasonable. However, in terms of shrinking $\underset{\sim}{X}$ towards a prior mean, this procedure is not altogether appealing. This choice of $N\underset{\sim}{\lambda}$ is as far from $\underset{\sim}{X}$ as possible, in terms of Euclidean distance; hence it is suggested the least by the data as the prior mean of $\underset{\sim}{X}$. Our formula for $\hat{k}(\underset{\sim}{\lambda})$ takes into account how far $\underset{\sim}{X}$ is from $N\underset{\sim}{\lambda}$, so that this defect is partially remedied. Thus, as long as the number of empty cells is relatively small, $\hat{k}_0$ will not be very large.

6. HANDLING BOTH FIXED AND RANDOM ZEROS

In this section we shall indicate how to extend our methods to adjust random zeros and simultaneously leave fixed zeros alone. The simplest of these extensions comes from the realization that shrinking $\underset{\sim}{X}$ toward the constant table only utilizes the multinomial nature of $\underset{\sim}{X}$; i.e., it ignores the cross-classification structure. In this sense, a fixed zero just means one less category.

Let f denote the number of fixed zeros. Then the effective number of categories is rc-f. Shrinking $\underset{\sim}{X}$ toward the constant table means choosing $\underset{\sim}{\lambda}$ so that $\lambda_{ij} = 0$ if the (i,j) cell corresponds to a fixed zero, and $\lambda_{ij} = 1/(rc-f)$ otherwise. In this case the expression for $\hat{k}(\underset{\sim}{\lambda})$, (4,3), reduces to

$$\hat{k}_2 = \frac{N^2 - \|\underset{\sim}{X}\|^2}{\|\underset{\sim}{X}\|^2 - N^2/(rc-f)}, \tag{6.1}$$

which is the same as $\hat{k}_1$ given in (4.5), except that rc is replaced by rc-f. Once again we note that this procedure alters the marginal totals. If margin preservation is desired, we can proceed in a manner similar to the second method of Section 4.

To illustrate this second approach, suppose that in Table 1 we wish to adjust the zero in the upper left hand corner but keep the other zero fixed. We must first choose an interaction structure for $\underset{\sim}{\lambda}$. Suppose we choose

$$\begin{aligned} \lambda_{ij} &= 0 \qquad \text{if cell } (i,j) \text{ has a fixed zero} \\ \lambda_{ij} &= a_i b_j \qquad \text{otherwise.} \end{aligned} \tag{6.2}$$

This is the "quasi-independence" model discussed in [6] and [19]. There are many tables satisfying (6.2). One of these is given by Table 6a.

TABLE 6

(a) Example of a table satisfying the model of quasi-independence.

(b) Final result of shrinking Table 1 towards a table with the interaction structure of Table 6a.

(a)

1	1	1	3
1	1	1	3
1	1	0	2
3	3	2	8

(b)

0.50	2.87	4.63	8.00
2.91	15.72	25.37	44.00
4.59	25.41	0	30.00
8.00	44.00	30.00	82.00

We use the iterative scaling technique described above in Table 6a to produce a new table which satisfies (6.2) and has the proper marginal totals. Because the iterations only involve row and column multiplication, the resulting table as $N\underset{\sim}{\lambda}$, we find that $\hat{k}(\underset{\sim}{\lambda})$ is a gigantic 1604. Shrinking Table 1 toward $N\underset{\sim}{\lambda}$ yields Table 6b which has the fixed zero but not the random zero. The large value of $\hat{k}(\underset{\sim}{\lambda})$, 1604, indicates that (6.2) is very much like the interaction structure in $\underset{\sim}{X}$ itself. Thus we find that Table 6b is very much like Table 1.

The technique used in the last example of smoothing a table can be generalized to allow for interaction structures that are more complicated than that implied by (6.2) This is achieved by still requiring $\lambda_{ij} = 0$ for all fixed zeros, but choosing the remaining λ's to give whatever interaction structure is deemed appropriate. Then the iterative scaling technique can be used to give the resulting $N\underset{\sim}{\lambda}$ the marginal totals of $\underset{\sim}{X}$. The resulting estimator will preserve both the margins of $\underset{\sim}{X}$ and all fixed zeros.

Now, a note of caution is necessary. In order for the Deming-Stephan procedure to work properly in the presence of fixed zeros, we must restrict the number and the positions of both the fixed and the random zeros. Suppose, for example, that we are given Table 7a to smooth.

TABLE 7

(a) Example of a table that cannot have certain zeros eliminated using the quasi-independent table given in Table 7b.

(b) A quasi-independent table.

(a)

0	2	0	2
2	0	2	4
0	2	0	2
2	4	2	8

(b)

0	1	1	2
1	0	1	2
1	1	0	2
2	2	2	6

Suppose also that the three diagonal zeros are fixed and the other two zeros are random. The second procedure used in this section would tell us to adjust Table 7b using the iterative scaling procedure so that it will have the marginal totals of Table 7a. Unfortunately, when this is done, we end up with Table 7a once again, and it still has the two zeros which we are trying to adjust.

When the $\underset{\sim}{\lambda}$ table is chosen to satisfy the quasi-independence model given by (6.2), and it is wished to adjust the margins so that they are the same as those of the given table, the following conditions (adapted from [6]) should be checked:

(a) the number of fixed zeros plus the number of random zeros is less than $(r-1)(c-1)$,

(b) each row and column of the $\underset{\sim}{\lambda}$ table has at lease one nonzero entry,

(c) the nonzero entries of the $\underset{\sim}{X}$ table form an inseparable table (i.e., the nonzero entries, after permutation of rows and columns, cannot be separated into two or more subtables, where each subtable has no row or columns in common with any other subtable).

These conditions are sufficient, although not always necessary for the iterative procedure to converge and to the desired values. Corresponding conditions for use when the $\underset{\sim}{\lambda}$ table is not quasi-independent vary greatly, and we suggest checking the table produced by the iterative procedure to see if it has zeros only in those cells corresponding to fixed zeros.

7. ASYMPTOTIC PROPERTIES OF THE ESTIMATORS

We have been proposing estimators of the form (4.4), for various values of $\underset{\sim}{\lambda}$, because of our desire for a reasonable solution to the problem of adjusting zero counts in sample contingency tables. But even the simplest such estimator, where we shrink $\underset{\sim}{X}$ toward the constant table, is a complicated nonlinear function of the cell counts, X_{ij}. As a result the exact sampling theory for these estimators can be found only by use of a computer. In this section we summarize some facts about the behavior of our estimates as the sample size, N, tends to infinity. Then, in the next section, we shall report some results on an exact evaluation of the risk function of the estimator where we shrink $\underset{\sim}{X}$ toward the constant table, and we compare it with the risk function of the raw table of counts, $\underset{\sim}{X}$.

In our asymptotic analysis we denote our choice of $\underset{\sim}{\lambda}$ as $\underset{\sim}{\hat{\lambda}} = \underset{\sim}{\hat{\lambda}}(\underset{\sim}{X})$, because, in general, we let $\underset{\sim}{\lambda}$ depend on $\underset{\sim}{X}$. To simplify this analysis we shall assume that there exists some $\underset{\sim}{\lambda}^* = \underset{\sim}{\lambda}^*(\underset{\sim}{p})$, such that $\sqrt{N}\,(\underset{\sim}{\hat{\lambda}}-\underset{\sim}{\lambda}^*)$ has an asymptotic, possibly degenerate, multivariate normal distribution with zero mean. For the simple case, where we shrink X toward the constant table, $\hat{\lambda}_{ij}(\underset{\sim}{X}) \equiv 1/rc \equiv \lambda^*_{ij}$, and thus $\sqrt{N}\,(\underset{\sim}{\hat{\lambda}}-\underset{\sim}{\lambda}^*) \equiv 0$, which is a degenerate case of the multivariate normal distribution. When we shrink X toward its independent projection, $\lambda_{ij}(\underset{\sim}{X}) = X_{i+}X_{+j}/N^2$, and if $\lambda^*_{ij}(\underset{\sim}{p}) = p_{i+}p_{+j}$, it is well known that $\sqrt{N}\ (\hat{\underset{\sim}{\lambda}}-\underset{\sim}{\lambda}^*)$ has an asymptotic multivariate normal distribution.

To simplify notation we set $\hat{\underset{\sim}{p}} = \underset{\sim}{X}/N$ and $\hat{w}_N = N/[N+\hat{k}(\underset{\sim}{\lambda})]$. Then we can rewrite the estimator given by (4.4) as

$$\underset{\sim}{B} = N[\hat{w}_N\hat{\underset{\sim}{p}}+(1-\hat{w}_n)\hat{\underset{\sim}{\lambda}}], \qquad (7.1)$$

and its risk function as

$$R(\underset{\sim}{p};\underset{\sim}{B}) = NE\|\hat{w}_N\sqrt{N}(\hat{\underset{\sim}{p}}-\underset{\sim}{p}) + (1-\hat{w}_N)\sqrt{N}(\hat{\underset{\sim}{\lambda}}-\underset{\sim}{p})\|^2. \qquad (7.2)$$

Since the risk of X is

$$R(\underset{\sim}{p};\underset{\sim}{X}) = N(1-\|\underset{\sim}{p}\|^2), \qquad (7.3)$$

the ratio of the risk of $\underset{\sim}{B}$ to that of $\underset{\sim}{X}$ is

$$\frac{E\|\hat{w}_N\sqrt{N}(\hat{\underset{\sim}{p}}-\underset{\sim}{p}) + (1-\hat{w}_N)\sqrt{N}(\hat{\underset{\sim}{\lambda}}-\underset{\sim}{p})\|^2}{1-\|\underset{\sim}{p}\|^2} \qquad (7.4)$$

Using the notation of this section, we can now rewrite $\hat{k} = \hat{k}(\underset{\sim}{\lambda})$ as

$$\hat{k} = \frac{1-\|\hat{\underset{\sim}{p}}\|^2}{\|\underset{\sim}{\lambda}-\hat{\underset{\sim}{p}}\|^2}. \qquad (7.5)$$

Since $\hat{\underset{\sim}{\lambda}}$ converges to $\underset{\sim}{\lambda}^*$ in probability, the following lemma is true.

<u>Lemma 1</u>: If $\underset{\sim}{p} \neq \underset{\sim}{\lambda}^*$, then $\hat{k}$ converges in probability to $(1-\|\underset{\sim}{p}\|^2)/\|\underset{\sim}{\lambda}^*-\underset{\sim}{p}\|^2$, and consequently $1-\hat{w}_N = 0_p(N^{-1})$.

Now we let $\underset{\sim}{U}_N = \sqrt{N}(\hat{\underset{\sim}{\lambda}}-\underset{\sim}{\lambda}^*)$ and $\underset{\sim}{V}_N = \sqrt{N}(\hat{\underset{\sim}{p}}-\underset{\sim}{p})$, and we assume that $(\underset{\sim}{U}_N,\underset{\sim}{V}_N)$ converges in distribution to some (U,V) having a multivariate normal distribution (which is possibly degenerate). The following result complements Lemma 1 for the case $\underset{\sim}{p} = \underset{\sim}{\lambda}^*$, and its simple proof is omitted.

<u>Lemma 2</u>: If $\underset{\sim}{p} = \underset{\sim}{\lambda}^*$, then $\hat{w}_N$ converges in distribution to the random variable

$$w = \frac{\|\underset{\sim}{U}-\underset{\sim}{V}\|^2}{1-\|\underset{\sim}{p}\|^2 + \|\underset{\sim}{U}-\underset{\sim}{V}\|^2}$$

To illustrate the implications of Lemma 2, let us look at the case when we shrink X toward the constant table. Since $\hat{\lambda}_{ij} \equiv \lambda^*_{ij} \equiv 1/rc$, $\underset{\sim}{U}_N \equiv 0$, so that $\underset{\sim}{U} \equiv 0$. If $p_{ij} = 1/rc$, then w simplifies to

$$w = \frac{\|\underset{\sim}{V}\|^2}{\|\underset{\sim}{V}\|^2+1-\|\underset{\sim}{p}\|^2}, \qquad (7.6)$$

where $\underset{\sim}{V}$ has a multivariate normal distribution with zero mean and covariance matrix of rank rc-1, given by

$$\begin{aligned} \mathrm{Var}(V_{ij}) &= p_{ij}(1-p_{ij}). \\ \mathrm{Cov}(V_{ij},V_{mn}) &= -p_{ij}p_{mn}, \text{ if } (i,j) \neq (m,n). \end{aligned} \tag{7.7}$$

A standard orthogonality argument shows that the distribution of w is the same as that of

$$\frac{C^2}{C^2+rc-1}$$

where C^2 has a chi-square distribution on rc-1 degrees of freedom. We shall use this fact shortly.

We now make use of our lemmas to show that the estimator,

$$\underset{\sim}{T} = \underset{\sim}{B}/N = \hat{w}_N\hat{\underset{\sim}{p}}+(1-\hat{w}_N)\hat{\underset{\sim}{\lambda}}, \tag{7.8}$$

is asymptotically equivalent to $\hat{p}$.

Theorem 2: (a) If $\underset{\sim}{p} \neq \lambda^*$, then $\sqrt{N}(\underset{\sim}{T}-\hat{\underset{\sim}{p}}) = 0_p(1/\sqrt{N})$.

(b) If $\underset{\sim}{p} = \underset{\sim}{\lambda}^*$, then $\sqrt{N}(\underset{\sim}{T}-\hat{\underset{\sim}{p}})$ converges in distribution to $(1-w)(\underset{\sim}{U}-\underset{\sim}{V})$ where w is given by (7.6).

Proof: We observe that, in general, $\sqrt{N}(\underset{\sim}{T}-\hat{\underset{\sim}{p}})$ may be written as

$$\begin{aligned} &(1-\hat{w}_N)[\sqrt{N}(\hat{\underset{\sim}{\lambda}}-\underset{\sim}{\lambda}^*) - \sqrt{N}(\hat{\underset{\sim}{p}}-\underset{\sim}{p}) + \sqrt{N}(\underset{\sim}{\lambda}^*-\underset{\sim}{p})] \\ &= (1-\hat{w}_N)[\underset{\sim}{U}_N-\underset{\sim}{V}_N + \sqrt{N}(\underset{\sim}{\lambda}^*-\underset{\sim}{p})]. \end{aligned} \tag{7.9}$$

In case (a), from Lemma 1 we see that $1-\hat{w}_N = 0_p(N^{-1})$, while $\underset{\sim}{U}_N,\underset{\sim}{V}_N$ and $\sqrt{N}(\underset{\sim}{\lambda}^*-\underset{\sim}{p})$ are all $0_p(\sqrt{N})$. Hence $\sqrt{N}(T-\hat{p}) = 0_p(1/\sqrt{N})$ when $\underset{\sim}{p} \neq \lambda^*$. In case (b), (7.9) reduces to

$$(1-\hat{w}_N)(\underset{\sim}{U}_N-\underset{\sim}{V}_N), \tag{7.10}$$

which converges in distribution to $(1-w)(\underset{\sim}{U}-\underset{\sim}{V})$.

Corollary: $\underset{\sim}{T}$ is a consistent estimator of $\underset{\sim}{p}$.

From part (a) of Theorem 2 we see that, if $\underset{\sim}{p} \neq \underset{\sim}{\lambda}^*$, $\underset{\sim}{T}$ is asymptotically equivalent to $\hat{\underset{\sim}{p}}$. Hence the ratio of the risk of $\underset{\sim}{T}$ to that of $\hat{\underset{\sim}{p}}$, given by (7.4), approaches one as $N\to\infty$.

When $\underset{\sim}{p} = \underset{\sim}{\lambda}^*$, the ratio of risks (7.4) converges to the value

$$\frac{E\|w\underset{\sim}{V}+(1-w)\underset{\sim}{U}\|^2}{1-\|\underset{\sim}{p}\|^2}\ . \tag{7.11}$$

When $\lambda_{ij}(\underset{\sim}{X}) \equiv 1/rc$ we can get some idea of the value of (7.11). In this case $\underset{\sim}{U} \equiv 0$, and if $p_{ij} \equiv 1/rc$, $1-\|p\|^2 = 1-1/rc$. Using (7.6) we can now rewrite (7.11) as

$$E\left[\left(\frac{rc\|\underset{\sim}{V}\|^2}{rc\|\underset{\sim}{V}\|^2 + rc-1}\right)^2 \cdot \frac{rc\|V\|^2}{rc-1}\right]. \tag{7.12}$$

But using the orthogonality argument mentioned earlier, it can be shown that (7.12) is equivalent to

$$E\left[\left(\frac{C^2}{C^2 + rc-1}\right)^2 \frac{C^2}{rc-1}\right], \tag{7.13}$$

where C^2 has a chi-square distribution on rc-1 degrees of freedom. Since

$$f(C^2) = \left(\frac{C^2}{C^2+rc-1}\right)^2 \frac{C^2}{rc-1} \tag{7.14}$$

is a convex function in C^2, by application of Jensen's inequality (see [7]) we see that

$$Ef(C^2) \geq f[E(C^2)] = 1/4. \tag{7.15}$$

Now, if rc-1 is large, $C^2/(rc-1)$ is approximately unity, consequently for large tables $E[f(C^2)]$ will be close to its expected value, 1/4. In the next section we investigate how close this result holds for small samples and small tables.

8. SMALL SAMPLE PROPERTIES OF THE ESTIMATORS

We denote by $\underset{\sim}{\Gamma} = \underset{\sim}{\Gamma}(\underset{\sim}{X})$ the estimator which shrinks $\underset{\sim}{X}$ toward the constant table, i.e.,

$$\Gamma_{ij} = \frac{N}{N+\hat{k}_1} X_{ij} + \frac{\hat{k}_1}{N+\hat{k}_1} \frac{N}{rc} \tag{8.1}$$

where

$$\hat{k}_1 = \frac{N^2 - \sum_{i,j} X_{ij}^2}{\sum_{i,j} X_{ij}^2 - N^2/rc} \tag{8.2}$$

We now report the results of preliminary investigations regarding the exact value of the risk function of $\underset{\sim}{\Gamma}$ which were carried out on the Harvard Time-sharing System (SDS-940).

Earlier we remarked that $\underset{\sim}{\Gamma}$ does not depend on $\underset{\sim}{X}$ being a cross-classification. Thus it is sensible for us to let the number of cells be 2, 3, and 4, although only the last number corresponds to an actual contingency table. In this section, therefore, we interpret $\underset{\sim}{\Gamma}$ as an estimator of a multinomial parameter $\underset{\sim}{p} = (p_1,\ldots,p_t)$ where t takes the values 2, 3, and 4. In graphs 1 through 6 we compare the risk functions of $\underset{\sim}{\Gamma}$ and $\underset{\sim}{X}$ for t = 2, 3, and 4, where the sample size N equals 15. We discuss each case in turn.

Graph 1 shows the risks functions of $\underset{\sim}{\Gamma}$ and $\underset{\sim}{X}$, where N = 15 and t = 2 (i.e., the binomial case). Here $\underset{\sim}{\Gamma}$ has smaller risk than $\underset{\sim}{X}$ for p_1 between 0.33 and 0.67; however, because $\underset{\sim}{X}$ is an admissible estimator of $\underset{\sim}{p} = (p_1,p_2)$, the risk of $\underset{\sim}{\Gamma}$ cannot be uniformly smaller than that of $\underset{\sim}{X}$. Thus, we see the "ears" in the risk function of $\underset{\sim}{\Gamma}$, which rise above the risk function of $\underset{\sim}{X}$. The asymptotic results of Section 7 indicate that, for $p_1 \neq 1/2$, as $N\to\infty$, these ears move toward each other and the difference between the two functions disappears. For $p_1 = 1/2$ the difference does not disappear. For N = 15, at $p_1 = 1/2$, the ratio of the risk of $\underset{\sim}{\Gamma}$ to the risk of $\underset{\sim}{X}$ is 0.51, a value which is considerably greater than the asymptotic lower bound of 1/4 discussed in the previous section.

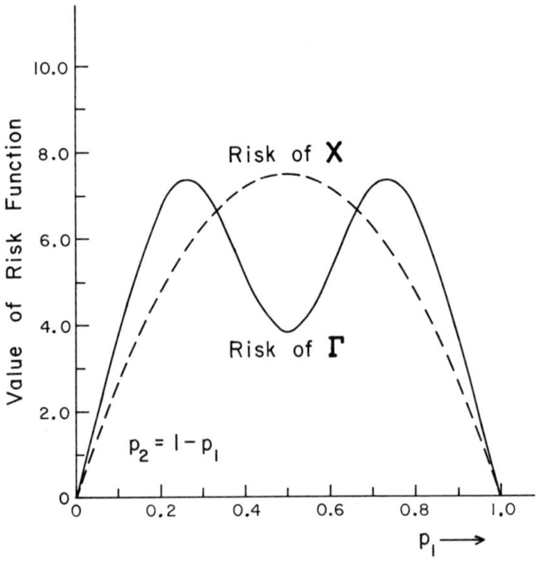

Graph 1. Risk Functions of $\underset{\sim}{X}$ and $\underset{\sim}{\Gamma}$ in the Binomial Case, N = 15.

We conjecture that the improvements in risk by using $\underset{\sim}{\Gamma}$ instead of $\underset{\sim}{X}$ accrue as the dimension of the multinomial increases. Thus we expect the binomial case to show $\underset{\sim}{\Gamma}$ at its worst. Although $\underset{\sim}{\Gamma}$ is superior to $\underset{\sim}{X}$ near $p_1 = 1/2$, for p_1 between 0.15 and 0.25 $\underset{\sim}{\Gamma}$ looks quite poor when compared with $\underset{\sim}{X}$.

Without resorting to three-dimensional pictures we cannot present the complete risk functions of $\underset{\sim}{\Gamma}$ and $\underset{\sim}{X}$ for t = 3. Instead of looking at these risk functions as they sit over the two-dimensional probability simplex, S_3 (see Section 4), we examine "sections" of them along two lines in S_3. Graph 2 shows the risk functions for those values of $\underset{\sim}{p}$ with $p_2 = p_3 = (1-p_1)/2$. Thus the "x-axis" gives the value of p_1 from 0 to 1, and the "y-axis" gives the value of the risk functions at the points $[p_1, (1-p_1)/2, (1-p_1)/2]$. By symmetry the picture is the same for points of the form $[(1-p_2)/2, p_2, (1-p_2)/2]$ or $[(1-p_3)/2, (1-p_3)/2, p_3]$. Graph 3 is similar to Graph 2 except p_3 is fixed at zero, so that $p_2 = 1-p_1$. Then the "y-axis" of Graph 3 gives the values of the risk functions at the points $(p_1, 1-p_1, 0)$. Again by symmetry we get the same picture for points of the form $(p_1, 0, 1-p_1)$ or $(0, p_2, 1-p_2)$.

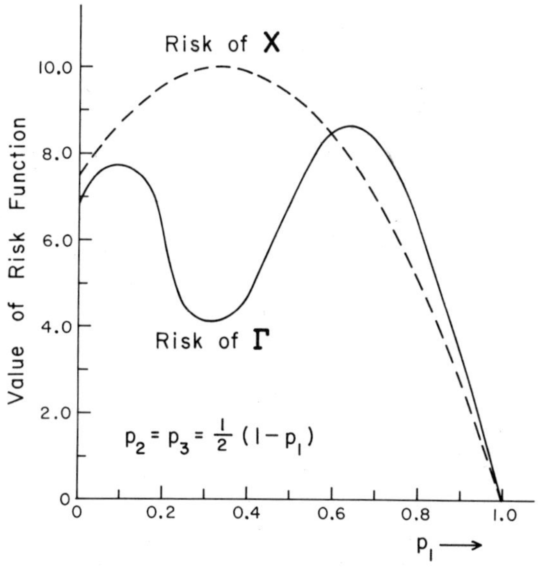

Graph 2. Section of Risk Functions of $\underset{\sim}{X}$ and $\underset{\sim}{\Gamma}$ in the Trinomial Case, N = 15.

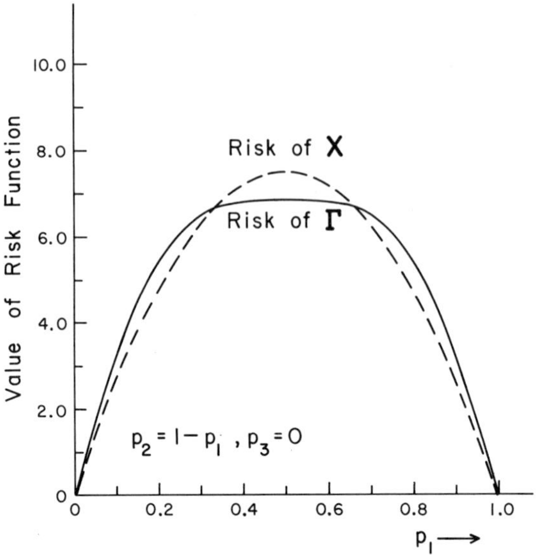

Graph 3. Section of Risk Functions of $\underset{\sim}{X}$ and $\underset{\sim}{\Gamma}$ in the Trinomial Case, N = 15.

Graph 3 indicates that for values of $\underset{\sim}{p}$ far away from the center of the simplex, (1/3,1/3,1/3), the risk functions of $\underset{\sim}{\Gamma}$ and $\underset{\sim}{X}$ are quite similar. This is in agreement with the asymptotic theory of the previous section. Graph 2 shows that, near the center of the simplex, $\underset{\sim}{\Gamma}$ has much smaller risk than $\underset{\sim}{X}$, while when p_1 exceeds 0.6, $\underset{\sim}{X}$ has the smaller risk. Although $\underset{\sim}{X}$ is somewhat better than $\underset{\sim}{\Gamma}$ when $p_1 > .6$, the difference between the two risk functions is less than the corresponding difference in the binomial case. At (1/3,1/3,1/3), the center of the simplex, the ratio of the risks is 0.42. This value is still larger than the asymptotic lower bound of 1/4, but it is closer to this value than in the binomial case.

For t = 4, we once again display "sections" of the risk functions for various lines through the simplex, S_4. Graph 4 corresponds to the section with $p_2 = p_3 = p_4 = (1-p_1)/3$. Graph 5 corresponds to the section with $p_4 = 0$ and $p_2 = p_3 = (1-p_1)/2$. Graph 6 corresponds to the section with $p_3 = p_4 = 0$ and $p_2 = 1-p_1$. By symmetry we can automatically get other sections of the simplex.

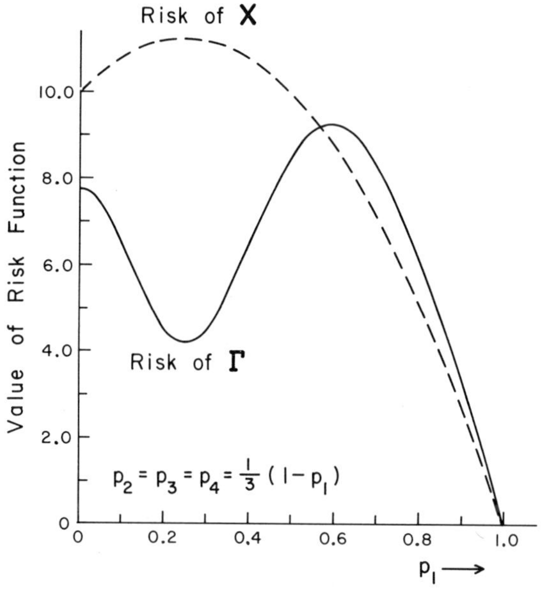

Graph 4. Section of Risk Functions of $\underset{\sim}{X}$ and $\underset{\sim}{\Gamma}$ in the Quadrinomial Case, N = 15.

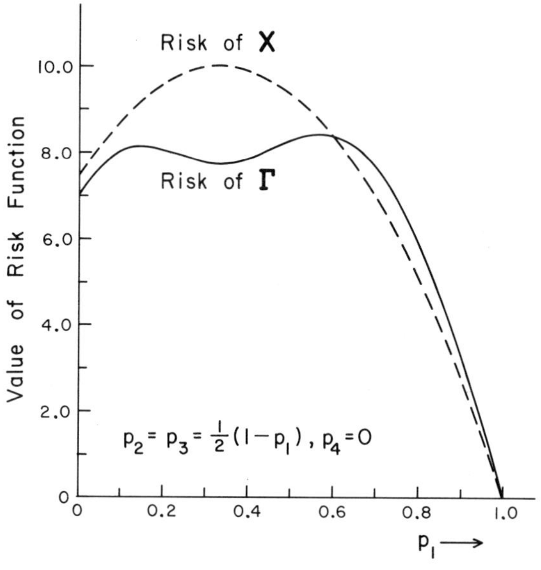

Graph 5. Section of Risk Functions of $\underset{\sim}{X}$ and $\underset{\sim}{\Gamma}$ in the Quadrinomial Case, N = 15.

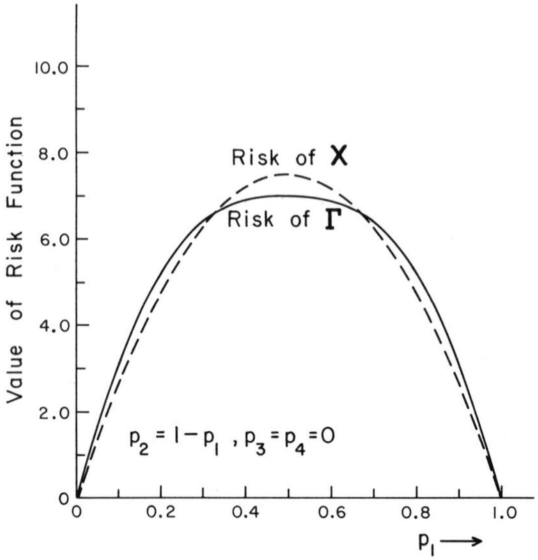

Graph 6. Section of Risk Functions of $\underset{\sim}{X}$ and $\underset{\sim}{\Gamma}$ in the Quadrinomial Case, N = 15.

As in the trinomial case, Graphs 5 and 6 show that for values of p far away from the center of the simplex, (1/4,1/4,1/4,1/4), the two risk functions as essentially equal. Graph 4 once again shows the substantial improvement of $\underset{\sim}{\Gamma}$ over $\underset{\sim}{X}$ for values of $\underset{\sim}{p}$ near the center. If no cell probability exceeds 0.56, and the remaining probabilities are of equal size, then $\underset{\sim}{\Gamma}$ is superior to $\underset{\sim}{X}$ in the sense of smaller mean-square error. When one of the probabilities does exceed 0.56, $\underset{\sim}{\Gamma}$ has somewhat larger risk than $\underset{\sim}{X}$, although the difference is not as great as the corresponding difference in the trinomial case.

The asymptotic theory indicates that as $N \to \infty$ the "ears" in Graphs 4 and 5 move toward the center of S_4, and the two risk functions will be essentially the same for $p \neq (1/4,1/4,1/4,1/4)$. At the center the ratio of the risk of $\underset{\sim}{\Gamma}$ to that of X, is 0.38, which is closer still to the asymptotic lower bound of 1/4 than the two previous cases.

In summary, these results indicate that $\underset{\sim}{\Gamma}$ is a viable competitor to the raw table, $\underset{\sim}{X}$, as an estimator of the vector parameter, $\underset{\sim}{p}$. The indications are that, for a fairly large neighborhood of parameter values around the center of the simplex, $\underset{\sim}{\Gamma}$ has much smaller risk than $\underset{\sim}{X}$. Furthermore, where $\underset{\sim}{\Gamma}$ does have larger risk than $\underset{\sim}{X}$, the difference is not very great. Finally, as the number of dimensions increases, i.e., the number of cell increases, the performance of $\underset{\sim}{\Gamma}$ relative to $\underset{\sim}{X}$ improves. These facts suggest that the asymptotic theory of real relevance to this problem is the one in which the number of cells, as well as the sample size, increases. If, in addition, the expected number of counts per cell remains finite, then the resulting asymptotic theory is clearly more relevant to the problem of sparcely filled contingency tables than the more standard theory of Section 7. Morris (see [24] and [25]) has considered asymptotic results of this sort, but he was concerned with problems of testing hypotheses.

9. FURTHER WORK

The results of Section 7 and 8 suggest the desirability of a more detailed analysis of the risk functions for the estimators which we suggest. In particular, it would be useful to determine the correctness of the conjecture that $\underset{\sim}{\Gamma}$ improves relative to $\underset{\sim}{X}$ as the number of cells increases. There is the possibility that the apparent improvement in Section 8 is an artifact of the particular value of N which we used. In general, estimators of the form (4.4) will have smaller risk than $\underset{\sim}{X}$ in a neighborhood of the region toward which $B(\hat{\underset{\sim}{\lambda}};\underset{\sim}{X})$ shrinks $\underset{\sim}{X}$. As yet we do not know the size of these neighborhoods, but the present evidence suggests they can be relatively large.

In estimating the optimal value of k, $\hat{k}(\underset{\sim}{\lambda})$, there are some alternatives which may yield a substantial improvement over our simple substitution of $\underset{\sim}{X}$ for $N\underset{\sim}{p}$. First we might estimate $\|\underset{\sim}{p}\|^2 = \sum_{i,j} p_{ij}^2$ in $\hat{k}_1$ by the sample "repeat rate",

$$\sum_{i,j} \frac{X_{ij}}{N}\left(\frac{X_{ij}-1}{N-1}\right),$$

as suggested in [15] and [16]. This is an unbiased estimate of $\|\underset{\sim}{p}\|^2$, whereas our estimate, $\|\underset{\sim}{X}\|^2$, is biased. Secondly, we might be better off estimating the weighting factor $w_N = N/(N+k)$ directly, rather than first estimating k and then substituting this to obtain $\hat{w}_N$. Thirdly, following the suggestion of

Mosteller [26], one may jackknife $\hat{w}_N$ to remove its first order bias and to obtain an estimate of its variability.

The analyses in this paper have been based on the quadratic loss function, $\|\underset{\sim}{T}(\underset{\sim}{X})-N\underset{\sim}{p}\|^2$. One can argue that this loss function does not pay enough attention to the very small cells expectations, which are the ones that really are of concern to us. Thus we might wish to use a loss function whose components are weighted inversely to the size of p_{ij}, e.g.,

$$\sum_{i,j} \frac{(T_{ij}-Np_{ij})^2}{Np_{ij}} .$$

The methods of estimation suggested here can easily be extended to higher dimensional tables. For example, in a three-way table we might use as $\hat{\underset{\sim}{\lambda}}$, the maximum likelihood estimator of $\underset{\sim}{p}$, under the hypothesis of no three-factor interaction (see [3]). Such an approach would have the nice property of allowing our estimator of $N\underset{\sim}{p}$ to retain some, but not an unwarranted amount, of the interaction exhibited by $\underset{\sim}{X}$. For higher-dimension tables and other interaction hypotheses, we would use similar procedures. The work of Bishop [1967, 1969] enables us to compute the expected cell values (our values for $N\underset{\sim}{\lambda}$) for multidimensional tables under a wide variety of interaction hypotheses.

ACKNOWLEDGEMENTS

We wish to thank Frederick Mosteller for many helpful comments and suggestions which aided in the preparation of this paper.

REFERENCES

[1] Bartlett, M. S. 1936. The square root transformation in analysis of variance. J. Roy. Statist. Soc. Suppl. 3:68.

[2] Blackwell, D., and Girshick, M. A. 1954. Theory of Games and Statistical Decisions. Wiley, New York.

[3] Birch, M. W. 1963. Maximum likelihood in three-way contingency tables. J. Roy. Statist. Soc., B. 25:220-233.

[4] Bishop, Y. M. M. 1967. Multidimensional Contingency Tables: Cell Estimates. Unpublished doctoral dissertation, Department of Statistics, Harvard University, Cambridge, Massachusetts.

[5] ————. 1969. Full contingency tables, logits, and split contingency tables. Biometrics. 24.

[6] ————., and Fienberg, S. E. 1969. Incomplete two-dimensional contingency tables. Biometrics. 25:119-28.

[7] Breiman, Leo 1968. Probability. Addison-Wesley, Reading, Mass., p. 80.

[8] Cochran, W. G. 1940. The analysis of variance when experimental errors follow the Poisson or Binomial laws. Ann. Math. Statist. 11:335-47.

[9] Deming, W. E., and Stephan, F. F. 1940. On a least squares adjustment of a sampling frequency table when the expected marginal totals are known. Ann. Math. Statist. 11:427-44.

[10] Fienberg, S. E. 1968. The geometry of an rxc contingency table. Ann. Math. Statist. 39:1186-90.

[11] ————. 1969. An iterative procedure for estimation in contingency tables. Unpublished manuscript.

[12] Freeman, M. F. and Tukey, J. W. 1950. Transformations related to the angular and the square root. Ann. Math. Statist. 21:607-11.

[13] Gart, J. J. 1962. Approximate confidence limits for relative risks. J. Roy. Statist. Soc., B. 24:454-63.

[14] ————., and Zweifel, J. R. 1967. On the bias of various estimators of the logit and its variance with application to quantal bioassay. Biometrika. 54:181-7.

[15] Good, I. J. 1965. The Estimation of Probabilities: An Essay on Modern Bayesian Methods. M.I.T. Research Monograph, N. 30.

[16] ————. 1967. A Bayesian Significance Test for Multinomial Distributions. J. Roy Statist. Soc., B. 29:344-431.

[17] Goodman, L. A. 1964a. A short computer program for the analysis of transaction flows. Behavioral Science. 9:176-86.

[18] ————. 1964b. Simultaneous confidence limits for cross-product ratios in contingency tables. J. Roy. Statist. Soc., B. 26:86-102.

[19] ————. 1968. The analysis of cross-classified data: Independence, quasi-independence, and interaction in contingency tables with or without missing entries, J. Amer. Statist. Assoc. 63:1091-1131.

[20] Haldane, J. B. S. 1955. A problem in the significance of small numbers. Biometrika. 42:266-7.

[21] Kullback, S., and Ireland, C. T. 1968. Contingency tables with given marginals. Biometrika. 55:179-88.

[22] Lidstone, G. J. 1920. Note on the general case of the Bayes-Laplace formula for inductive or a posteriori probabilities. Trans. Fac. Actuar. 8:182-92.

[23] Lindley, D. V. 1964. The Bayesian analysis of contingency tables. Ann. Math. Statist. 35:1622-43.

[24] Morris, C. 1966. Admissible Bayes procedures and classes of epsilon Bayes procedures for testing hypotheses in a multinomial distribution. Technical Report No. 55, Nonr-225(72), Statistics Department, Stanford University.

[25] ——————. 1967. A class of epsilon Bayes tests of a simple null hypothesis on many parameters in exponential families. Technical Report No. 57, Nonr-225(72), Statistics Department, Stanford University.

[26] Mosteller, F. 1968. Association and estimation in contingency tables. J. Amer. Stat. Assoc. 63:1-28.

[27] Plackett, R. L. 1962. A note on interactions in contingency tables. J. Roy. Statist. Soc., B. 24:162-6.

ON FUNCTIONS OF DISCRETE MARKOV CHAINS

CHIA KUEI TSAO
Department of Mathematics
Wayne State University
Detroit, Michigan

SUMMARY

This article represents a generalization of results in a previous work [9] by the author. For a Markov chain with infinite state space, a chain is represented as a finite sequence of sequences of real numbers which possess similar properties as indicator random variables. The numbers of occurrences of various groups of states in the chain are counted by functions of such sequences, and a few examples are given.

1. INTRODUCTION

Some functions of Markov chains are useful in practical applications, and, for certain Markov chains, many such functions become linear combinations of independent random variables (or vectors) and thus possess easily obtainable distributions. A few results of this type for finite state Markov chains were derived in [9]. The method essentially involves the use of random vectors which serve some of the purposes of indicator random variables. The components of these vectors were used as counting devices for counting the numbers of occurrences of certain events in a chain. In some cases, these latter numbers become linear combinations of independent random variables.

The present paper is a generalization of [9] to the cases where the state space is infinite. As a generalization of [9], a state will be represented by an indicator-like sequence of real numbers, one of which is unity and all others are zero. Thus, if $z_1, z_2, z_3 \ldots$ denote the various states, then we write

$$\begin{aligned} z_1 &= (1,0,0,0,0,\ldots) \\ z_2 &= (0,1,0,0,0,\ldots) \\ z_3 &= (0,0,1,0,0,\ldots) \end{aligned} \tag{1.1}$$

Consequently, if we denote by $Z_0, Z_1, \ldots, Z_T$ the random sequence of states of a Markov chain for time $t = 0,1,2,\ldots,T$, then each Z will take on one of the z's defined in (1.1), and some functions of the chain can be expressed as linear combinations of products of consecutive Z's. For some Markov chains, these functions can be considered linear functions of independent random variables. This problem will be discussed in section 2, and a few examples will be given in section 3.

2. THE THEORY

Let $\mathcal{X} = \{z_1, z_2, \ldots\}$ be a state space consisting of the states $z_1, z_2, \ldots$, where each state is represented as a sequence of real numbers, of which one is unity and all the remaining ones are zero. More precisely, we have

$$z_i = (\delta_{i1}, \delta_{i2}, \ldots), \quad i = 1,2,\ldots \tag{2.1}$$

$$\text{where } \delta_{ij} = \begin{cases} 1 \text{ if } i = j \\ 0 \text{ if } i \neq j \end{cases} \tag{2.2}$$

Let t denote a time parameter which assumes the values 0,1,2,..., and let Z_t denote a random sequence at time t, which assumes values in $\mathcal{X}$; that is,

$$Z_t = (Z_{t1}, Z_{t2}, \ldots), \quad t = 0,1,2,\ldots, \tag{2.3}$$

where one of the random components $Z_{t1}, Z_{t2}, \ldots$ assumes the value unity and all other components take on the value zero.

Let $Z_0, Z_1, \ldots, Z_t$ be a Markov chain with an initial probability distribution Q and a stationary transition probability matrix P given by

$$Q = (q_1, q_2, \ldots) \tag{2.4}$$

$$P = \begin{Vmatrix} P_{11} & P_{12} & \cdots \\ P_{21} & P_{22} & \cdots \\ & \cdots\cdots & \\ & \cdots\cdots & \\ & \cdots\cdots & \end{Vmatrix} \tag{2.5}$$

That is, we have

$$q_i = \Pr Z_0 = z_i$$

$$= \exp\left[\sum_{j=1}^{\infty} \delta_{ij} \ln q_j\right], \quad i = 1,2,\ldots \tag{2.6}$$

$$P_{ij} = \Pr\left[Z_t = z_j | Z_0 = z_{i_0}, \ldots, Z_{y-1} = i\right]$$

$$= \Pr\left[Z_t = z_j | Z_{t-1} = i\right] \tag{2.7}$$

$$= \exp\left[\sum_{u=1}^{\infty} \sum_{v=1}^{\infty} \delta_{iu}\delta_{jv} \ln p_{uv}\right], \quad i,j = 1,2,\ldots.$$

In (2.6) and (2.7) we have defined

$$0\ (\ln 0) = 0$$

$$1\ (\ln 0) = -\infty, \tag{2.8}$$

so that (2.6) and (2.7) are equivalent to (2.4) and (2.5), respectively, even when certain elements p_{ij} are not positive. With the above notations, the probability distribution function of a Markov chain (that is, the probability of obtaining a particular chain $z_{i_0}, z_{i_1}, \ldots, z_{i_T}$) may be written as

$$f(z_{i_0}, \ldots, z_{i_T}) = \Pr\left[Z_0 = z_{i_0}, \ldots, Z_T = Z_{i_T}\right] \tag{2.9}$$

$$= \exp\left[\sum_{j=1}^{\infty} \delta_{i_0 j} \ln q_j + \sum_{t=1}^{T} \sum_{u=1}^{\infty} \sum_{v=1}^{\infty} \delta_{i_{t-1} u} \, \delta_{i_t v} \ln p_{uv}\right],$$

$i_t = 1,2,\ldots,\ t = 0,1,2,\ldots,T.$

Now, we shall use the following device for counting the numbers of occurrences of groups of states in the chain.

$$\text{Let } A = \begin{Vmatrix} a(1,1) & a(1,2) & \ldots \\ a(2,1) & a(2,2) & \ldots \\ & \ldots\ldots & \\ & \ldots\ldots & \\ & \ldots\ldots & \end{Vmatrix} \qquad (2.10)$$

be a matrix such that each row of A is a rearrangement of the positive integers 1,2,....

$$\text{Let } M = \begin{Vmatrix} m_{10} & m_{11} & m_{12} & \cdots \\ m_{20} & m_{21} & m_{22} & \cdots \\ & \ldots\ldots & & \\ & \ldots\ldots & & \\ & \ldots\ldots & & \end{Vmatrix} \qquad (2.11)$$

be a matrix of non-negative integers such that each row of M is a sequence of strictly increasing integers with $m_{u0} = 0$, $u = 1,2,\ldots$. Let $X_1, X_2, \ldots, X_T$ be T random sequences which are functions of $Z_0, \ldots, Z_T$, defined by

$$X_t = (X_{t1}, X_{t2}, \ldots),\ t = 1,2,\ldots,T, \qquad (2.12)$$

$$\text{where } X_{tj} = \sum_{u=1}^{\infty} \sum_{v=m_{u(j-1)}+1}^{m_{uj}} Z_{(t-1)u} Z_{ta(u,v)},\ j = 1,2,\ldots. \qquad (2.13)$$

It is obvious that $x_1, \ldots, X_T$ is a sequence of random sequences, each defined on the state space $\mathfrak{X}$. These random sequences are not generally stochastically independent. However, if the transition probability matrix $p = \|p_{ij}\|$ of the Markov chain $Z_0, \ldots, Z_T$ satisfies the property that for each u ($u = 1,2,\ldots$) and each j ($j = 1,2,\ldots$), then the set of transition probabilities

$$p_{ua(u,m_{uj-1}+1)}, \ldots, p_{ua(u,m_{uj}}$$

is a scalar multiple of a m_{uj}-variate multinomial distribution

$$\alpha_{uj} = (\alpha_{uj1}, \ldots, \alpha_{ujm_{uj}}). \qquad (2.14)$$

In other words, we have

$$p_{ua(u,m_{u(j-1)}+1)}, \ldots, p_{ua(u,m_{uj})} \qquad (2.15)$$

$$= p_j(\alpha_{uj1},\ldots,\alpha_{ujm_{uj}}) = p_j \alpha_{uj},$$

$$j = 1,2,\ldots;\ u = 1,2,\ldots,$$

where $(p_1,p_2,\ldots)$ is some probability distribution. Then $X_1,\ldots,X_T$ are independently and identically distributed random sequences with the common distribution $(p_1,p_2,\ldots)$; that is,

$$\Pr\left[X_t = z_i\right] = p_i, \text{ for } i = 1,2,\ldots, \text{ and } t = 1,2,\ldots,T. \quad (2.16)$$

Let $S_T = (s_{T1},s_{T2},\ldots)$ (2.17)

$$= (X_{11}+X_{21}+\ldots+X_{T1},X_{12}+X_{22}+\ldots+X_{T2},\ldots)$$

$$= X_1 + X_2 + \ldots + X_T.$$

Then the components of S_T represent the numbers of occurrences of the different groups of states in the chain. Let $c_1,c_2,\ldots$ be a sequence of real numbers, not all being equal, and let

$$R_T = c_1 s_{T1}+c_2 s_{T2}+ \ldots. \quad (2.18)$$

Then the distribution of R_T is given by

$$\Pr\left[R_T = k\right] = \sum p_1^{k_1} p_2^{k_2}\ldots, \quad (2.19)$$

where each summand is the product of a finite number of the p's and the summation $\sum$ is taken over all possible combinations of finite numbers of the k's such that $c_1k_1+c_2k_2+\ldots = k$.

In applications, various random variables may be obtained from R_T if one chooses different sequences $c_1,c_2,\ldots,$ and their distributions may be obtained from (2.19). A few simple examples will be given in the next section.

3. EXAMPLES

The random variable R_T defined in (2.18) represents a linear combination of the numbers of occurrences of the various groups of states in the Markov chain partitioned through the use of (2.10) and (2.11). In the case where $X_1,X_2,\ldots,X_T$ are independent, the distribution of R_T may be easily obtained from (2.19). For illustrations, the following are a few examples:

(a) If $c_1,c_2,\ldots$ are chosen such that for a fixed i

$$c_i = 1$$

$$c_j = 0, \quad j \neq i, \quad (3.1)$$

then the distribution of R_T becomes simply the binomial probability distribution

$$\Pr\left[R_T = k\right] = \Pr\left[S_{Ti} = k\right] \tag{3.2}$$
$$= \binom{T}{k} p_i{}^k (1-p_i)^{T-k}, \text{ where } k = 0,1,\ldots,T.$$

Example (recurrent events): Consider a special case of a Markov chain concerning recurrent events having the following transition probability matrix (see [4], pp. 344-5):

$$P = \left\| \begin{matrix} q & p & 0 & 0 & 0 & \ldots \\ q & 0 & p & 0 & 0 & \ldots \\ q & 0 & 0 & p & 0 & \ldots \\ & & \ldots\ldots & & & \\ & & \ldots\ldots & & & \\ & & \ldots\ldots & & & \end{matrix} \right\| . \tag{3.3}$$

$$\text{Let } A = \left\| \begin{matrix} 1 & 2 & 3 & 4 & \ldots \\ 1 & 3 & 2 & 4 & \ldots \\ 1 & 4 & 2 & 3 & \ldots \\ & & \ldots\ldots & & \\ & & \ldots\ldots & & \\ & & \ldots\ldots & & \end{matrix} \right\| . \tag{3.4}$$

$$M = \left\| \begin{matrix} 0 & 1 & 2 & 3 & \ldots \\ 0 & 1 & 2 & 3 & \ldots \\ 0 & 1 & 2 & 3 & \ldots \\ & & \ldots\ldots & & \\ & & \ldots\ldots & & \\ & & \ldots\ldots & & \end{matrix} \right\| . \tag{3.5}$$

Then $X_1, X_2, \ldots$ are binomial random vectors with the common distribution

$$\begin{aligned} \Pr\left[X_t = z_1\right] &= q, \quad t = 1,2,\ldots,T, \\ \Pr\left[X_t = z_2\right] &= p, \quad t = 1,2,\ldots,T, \\ \Pr\left[X_t = z_j\right] &= 0, \quad j = 3,4,\ldots, \quad t = 1,2,\ldots,T \end{aligned} \tag{3.6}$$

and $R_T = S_{T1}$ is the number of occurrences of state Z_1 (recurrent event) in the chain of length T; and

$$\Pr\ S_{T1} = k = \binom{T}{k} q^k p^{T-k}, \quad k = 0,1,2,\ldots,T. \tag{3.7}$$

We note that this random variable $R_T = S_{T1}$ can be interpreted as the number of occurrences of zero-waiting times (see [4], example (i), pp. 344-5), or as the number of occurrences of failures in a chain of "success runs" (example (h), p. 344), or as the number of moves in one direction of an unrestricted random walks (example (e), p. 343).

(b) If, in a Markov chain, we have

$$P = \begin{Vmatrix} p_1 & p_2 & \cdots \\ p_1 & p_2 & \cdots \\ & \cdots\cdots & \\ & \cdots\cdots & \\ & \cdots\cdots & \end{Vmatrix}, \tag{3.8}$$

that is, if the rows of p are identical distributions and if we let

$$A = \begin{Vmatrix} 1 & 2 & 3 & \cdots \\ 1 & 2 & 3 & \cdots \\ & \cdots\cdots & & \\ & \cdots\cdots & & \\ & \cdots\cdots & & \end{Vmatrix}, \tag{3.9}$$

$$M = \begin{Vmatrix} 0 & 1 & 2 & 3 & \cdots \\ 0 & 1 & 2 & 3 & \cdots \\ & \cdots\cdots & & & \\ & \cdots\cdots & & & \\ & \cdots\cdots & & & \end{Vmatrix}, \tag{3.10}$$

where $c_i = i-1$, and $i = 1,2,\ldots,$ (3.11)

then R_T becomes the sum of T independently, identically distributed random variables $Y_1, Y_2, \ldots, Y_T$ with the common distribution

$$\Pr\left[Y_i = k\right] = p_{k+1},\ k = 0,1,2,\ldots. \tag{3.12}$$

As an example, suppose

$$p_i = \frac{\mu^{i-1}e^{-\mu}}{(i-1)!},\ i = 1,2,\ldots. \tag{3.13}$$

$$\text{Then } \Pr = \left[R_T = k\right] = \frac{(T\mu)^k e^{-T\mu}}{k!},\ k = 0,1,\ldots. \tag{3.14}$$

That is, if one is taking a random sample of size T (omitting the initial value) from a Poisson distribution (considered as a Markov chain), then $S_{T1}, S_{T2}, \ldots$ represent the numbers of occurrences of the values $0,1,2,\ldots,$ and

$$R_T = S_{T2} + 2s_{T3} + 3s_{T4} + \cdots \tag{3.15}$$

is exactly the sum of the random sample of size T from Poisson distribution (3.13). Thus R_T, as defined by (3.15) is distributed by the Poisson distribution in (3.14).

REFERENCES

[1] Anderson, T. W., and Goodman, Leo A. 1957. Statistical inference about Markov chain. Ann. Math. Statist. 28:89-110.

[2] Barton, D. E.; David, F. N.; and Fix, E. 1962. Persistence in a chain of multiple events when there is simple dependence. Biometrika 49:351-7.

[3] Baum, L. E., and Petrie, T. 1966. Statistical inference for probabilistic functions of finite state Markov chains. Ann. Math. Statist. 37:1554-63.

[4] Feller, W. 1960. An Introduction to Probability Theory and Its Applications 2nd ed., rev. New York: John Wiley and Sons.

[5] Goodman, Leo A. 1958. Simplified run tests and likelihood ratio tests for Markov chains. Biometrika 45:181-97.

[6] Goodman, Leo A. 1964. The analysis of persistence in a chain of multiple events. Biometrika 51:405-11.

[7] Parzen, E. 1962. Stochastic Processes. San Francisco: Holden-Day.

[8] Prais, S. J. 1955. Measuring social mobility. J. Roy. Statist. Soc. Ser. A 118:56-66.

[9] Tsao, C. K. 1968. Admissibility and distribution of some probabilistic functions of discrete finite state Markov chains. Ann. Math. Statist. 39:1649-53.